DATE OF RETURN
UNLESS RECALLED BY LIBRARY

PLEASE TAKE GOOD CARE OF THIS BOOK

Pearson Education

We work with leading authors to develop the
strongest educational materials in geography,
bringing cutting-edge thinking and best learning
practice to a global market.

Under a range of well-known imprints, including
Prentice Hall, we craft high quality print and
electronic publications which help readers to
understand and apply their content,
whether studying or at work.

To find out more about the complete range of our
publishing, please visit us on the World Wide web at:
www.pearsoneduc.com

Dissident Geographies

An Introduction to Radical Ideas and Practice

Alison Blunt and Jane Wills

An imprint of **Pearson Education**

Harlow, England · London · New York · Reading, Massachusetts · San Francisco
Toronto · Don Mills, Ontario · Sydney · Tokyo · Singapore · Hong Kong · Seoul
Taipei · Cape Town · Madrid · Mexico City · Amsterdam · Munich · Paris · Milan

Pearson Education Limited
Edinburgh Gate
Harlow
Essex CM20 2JE
England

and Associated Companies throughout the world

Visit us on the world wide web at:
http://www.pearsoneduc.com

First published 2000

ISBN 0 582 29489 4

British Library Cataloguing in Publication Data
A catalogue record for this book is available from the British Library

Library of Congress Cataloging-in-Publication Data
A catalog record for this book is available from the Library of Congress

Typeset by No. 42

Produced by Pearson Education Asia Pte Ltd
Printed and bound by CPI Antony Rowe, Eastbourne

in memory of Tren Wills

Contents

Acknowledgements

Many people have helped to make this book possible. In particular, we would like to thank David Pinder for his enthusiastic support and for writing Box 1.10 in Chapter 1. Noel Castree, Derek Gregory and Linda McDowell all provided positive encouragement at the beginning of this project, and Noel has also made helpful comments on a draft. We have also benefited from the constructive criticism of Liz Fox, Kristie Legg and Richard Phillips, all of whom have read drafts of the text. We would also like to thank our students for helping us to crystallise these ideas through debate. We thank Danny Cusack, Joe Murray and everyone at AFrI for their warm welcome and inspiring ideas that had a profound impact on us and our students on several fieldtrips to Ireland. Sally Wilkinson commissioned this book and Matthew Smith and Shuet-kei Cheung at Pearson Education have seen it through to completion. Linda Hall has drawn the wonderful maps and Cynthia Norman helped with the production of the manuscript. Finally, writing *Dissident Geographies* has been made all the more pleasurable by the support of Cecily, Peter and David Blunt, Jim Chapman and Greta Wills, and by the friendship of Andrew Lincoln, Andy Merrifield, Elaine Sharland and Jane Tooke.

We would like to thank the following for permission to reproduce copyright material; David King Collection for figure 1.1 and 1.2; Joy Thacker for figure 1.3; REX Features for figure 1.6; Camera Press Ltd for figure 1.10.1; Roger-Violett for figure 1.10.2; Mary Evans Picture Library for figure 2.1; the National Museum of Labour History for figure 2.2; Corbis-Bettmann/UPI for figure 2.3; Dave Sinclair for figure 2.9.1; LIBERTY for figure 3.1; Mary Evans Picture Library/Fawcett Library for figure 3.2; the Museum of London for figure 3.3; Corbis-Bettmann/UPI for figure 3.4; Brenda Prince/FORMAT for figure 3.5; Corbis-Bettmann/UPI for figure 3.2.1; the Bodleian Library, Oxford for figure 4.1 from MS. Raul. D. 939 part 3 – detail; Jacky Fleming for figure 4.2; Corbis-Bettmann/UPI for figure 4.3; Costa Manos/MAGNUM Photos for figure 4.4; Fraser Hall/Robert Harding Picture Library for figure 4.5; Angela Martin for figure 4.6; Peter Newark's American Pictures for figure 5.1; Mary Evans Picture Library for figure 5.2; Joe Murray and Action From Ireland for figures 5.4 and 5.5; Camera Press for figure 5.2.1; Nicolas Tikhamiroff/MAGNUM Photos Inc for figure 5.4.1.

We have made every effort to trace copyright holders, but if we have inadvertently overlooked any, we will be pleased to rectify our error at the first opportunity.

Dissident Geographies:
An Introduction

This book is about radical ideas and practices, their geographical origins and manifestations, and their implications for geographical thought. The dissident geographies explored in the book all share a political commitment to overturning prevailing relations of power and oppression. Each chapter introduces a different strand of radical thought and action *before* going on to examine the contexts in which these ideas and practices developed and their geographical implications in more depth. The book has a threefold approach to dissident geographies; first, we introduce the spaces and places within which different radical ideas are produced and practised; second, we explore the impact these ideas have had on social and political landscapes; and third, we unpack the implications of these ideas for the scholarly discipline of geography. As such, *Dissident Geographies* explores the spatiality of political practice *and* the politics of geographical thought.

Dissident Geographies aims to introduce a range of radical ideas which have shaped, and continue to shape, the ways the world is understood, experienced and changed for the better. The book illustrates the ways in which these political traditions and activism outside the academy shape the production and dissemination of knowledge within geography, by tracing the disciplinary development and contribution of a number of dissident ideas. The radical traditions that we include are anarchism, marxism, feminism, the struggle for sexual liberation, and postcolonialism. Each of these has a distinct history and geography, different agendas for action and diverse implications for the contemporary world. But each of these radical traditions also has important and productive links to other bodies of thought and action, and their protagonists have often formed political alliances with one another. Each set of radical ideas has inspired new visions of past, present and future worlds, and each tradition has adherents who are drawing on those ideas to inform their behaviour, alliances and actions today. Moreover, the development of these ideas within the discipline of geography has shown that geographical knowledge is not – and should not attempt to be – static and detached from what is going on in the world, but is rather dynamic and profoundly influenced by events, struggles and politics beyond university life. If we are to understand and change the contemporary world, *Dissident Geographies* is designed to help illuminate the history and geography of radical ideas and so inform analysis and action today (for more on the broader history of geographical thought see Barnes and Gregory, 1997; Cloke *et al.*, 1991; Daniels and Lee, 1996; Gregory *et al.*, 1994; Livingstone, 1993; Massey *et al.*, 1999; Peet, 1998; Unwin, 1992. For more on activism and the academy see Blomley, 1994; Castree, 1999; Tickell, 1995).

Dissident Geographies examines the impact of different radical ideas in shaping the ways in which geography is researched, taught and institutionalised as an academic discipline. Each set of ideas that we examine has raised new research agendas and new research methods within geography while also highlighting pervasive power inequalities between geographers, both staff and students. Radical ideas thus have implications not only for *what* geographers do but also for *how* they do it. In putting new issues on the agenda, anarchist, marxist, feminist, sexual and postcolonial geographers propose new ways of relating to each other in the context of new institutional practices within the discipline. Less hierarchical, more inclusive, relations within the discipline are argued to be important in determining *who* is attracted to being a geographer, *what* they are able to do once they enter the discipline, and their ability to take up positions of power. Dissident geographers seek to overturn traditional power relations and attract working class people, women, sexual dissidents and people of colour to a discipline in which they can flourish and progress to play leadership roles.

Ignited by the wave of radical protests that spanned the Civil Rights and anti-Vietnam war movements, student unrest and the campaigns for gay and women's liberation, dissident geographers began to challenge the prevailing orthodoxies of the discipline in the late 1960s and early 1970s. As part of this endeavour, a group of graduate students and faculty members in the Geography Department at Clark University in the United States set up a new journal called *Antipode: A Radical Journal of Geography* in 1969. In the first issue, Richard Peet argued that radical geography posed important challenges for the future: first, it could help to design and campaign for a more equitable society 'in which poverty, suffering and the deadening feeling of uselessness and helplessness are eradicated, and in which a free people achieve a higher order of existence' (Peet, 1969: 4); second, it could help to achieve radical change through argument, polemic and activism; and third, it could help to restructure academic geography, to democratise its institutions, and to change what was taught. Radical geography was to be about designing and fighting for social change as well as revolutionising the discipline of geography.

In the early days, this awakening of radical geography was mainly characterised by geographical interpretations of anarchist and marxist thought, but subsequently, over the course of the 1980s and 1990s, the radical agenda has widened to include feminist, sexual liberationist and postcolonial geographies. Thirty years after the first issue of *Antipode* was published, this book examines the importance of radical ideas and practices in geography today. While there can be no doubt that dissident geographies have had a great impact on the discipline, there remains much more to be done. Inequality and injustice have not disappeared and, in many ways, they remain more pressing than ever. Dissident geographies remain crucially important in attempting to change both the discipline of geography and the world.

References

BARNES, T. and GREGORY, D. (eds) (1997) *Reading Human Geography: the poetics and politics of inquiry.* London: Arnold.

BLOMLEY, N. (1994) Activism and the academy. *Environment and Planning D: Society and Space*, **12**: 383–5.

CASTREE, N. (1999) Out there? In here? Domesticating critical geography. *Area*, **31**: 81–6.

CLOKE, P., PHILO, C. and SADLER, D. (1991) *Approaching Human Geography: an introduction to contemporary theoretical debates.* London: Paul Chapman.

DANIELS, S. and LEE, R. (eds) (1996) *Exploring Human Geography: a reader.* London: Arnold.

GREGORY, D., MARTIN, R. and SMITH, G. (eds) (1994) *Human Geography: society, space and social science.* Basingstoke: Macmillan.

LIVINGSTONE, D. (1993) *The Geographical Tradition: episodes in the history of a contested enterprise.* Oxford: Blackwell.

MASSEY, D., ALLEN, J. and SARRE, P. (eds) (1999) *Human Geography Today.* Cambridge: Polity.

PEET, R. (1969) A New Left geography. *Antipode*, **1**: 3–5.

PEET, R. (1998) *Modern Geographical Thought.* Oxford: Blackwell.

TICKELL, A. (1995) Reflections on activism and the academy. *Environment and Planning D: Society and Space*, **13**: 235–7.

UNWIN, T. (1992) *The Place of Geography.* Harlow: Longman.

1

The Fire of Liberty: Anarchism and Geography

Anarchism

The Greek word *anarchos* simply means 'without a ruler', and the word anarchy is often used to describe the social disorder, violence and chaos associated with the breakdown of authority and the widespread violation of law. Yet anarchists aspire to the absence of authority as a positive step on the road to building a new society in harmony with itself and with nature. Rather than being a negative term, anarchy is argued to be a positive social development, allowing each individual to blossom without the restrictions and confinements of authoritative power, law and control. By its nature, however, this tradition of dissent is eclectic and rather hard to pin down. Anarchist writers have tended to eschew definitive political programmes or organisational practices, and there has been little co-ordination between anarchist groups. Indeed, as Fauré suggests, anarchists are only really united in their opposition to authority in all its forms, and beyond that, there is enormous diversity within the tradition:

> There may be – and indeed there are – many varieties of anarchist, yet all have a common characteristic that separates them from the rest of humankind. This uniting point is *the negation of the principle of Authority in social organisations and the hatred of all constraints that originate in institutions fuelled on this principle.* Thus, whoever denies Authority and fights against it is an Anarchist. (Fauré, quoted in Woodcock, 1977: 62; emphasis in the original)

Despite its antecedents in all human rebellion, and particularly in the political battles of the English Civil War and the French Revolution, the anarchist tradition only came to self-consciousness in the mid-nineteenth century. Anarchist thinkers such as Pierre Joseph Proudhon, Michael Bakunin, Peter Kropotkin and Elisée Réclus were part of the wider socialist movement, and it was not until the 1870s that anarchists began to clearly distinguish themselves from marxists in arguments over the state, leadership and the mechanisms necessary to achieve social change. In this chapter we focus on the key ideas of these nineteenth-century anarchists (see Box 1.1), we explore the geography of anarchist organisation and experiment, and we consider the disciplinary implications of anarchist thought as far as geography is concerned. In this regard, it is significant that two of the key protagonists in the history of anarchism were practising geographers. A profound interest in the environment and in the diversity of social formations inspired both the geography and the anarchism of Peter Kropotkin and Elisée Réclus, and in their day, both men were celebrated as scholars of physical and regional geography (see Box 1.2 for details of the life of Kropotkin and Box 1.3 for a summary of Réclus's life and work and his involvement in the Paris Commune of 1871). For our purposes, however, it is frustrating that Kropotkin and Réclus were not able to combine their anarchist ideas with their geographical scholarship as they might do today. Moreover, although a number of authors have sought to spell out the geographical implications of the anarchist writings of Kropotkin and Réclus (see Stoddart, 1975; Galois, 1976; Breitbart, 1975, 1981; Dunbar, 1978, 1981; Fleming, 1988; Cook and Pepper, 1990), there has been little development of anarchism in geographical theory and/or research, leaving us to speculate about what an anarchist geography might be like. To date, anarchism has made its clearest mark on geography by influencing a new generation of academics in the late 1960s and 1970s, inspiring them to question the authority, hierarchies and received wisdoms of the discipline. Such anarchist-inspired rebellion brought forth the new shoots of a radical geography associated with the journal *Antipode*, the development of new research themes, new disciplinary practices and the breakthrough to marxism discussed in Chapter 2. Anarchist ideas have inspired enormous change within the discipline, but as yet, they have spawned only the outlines of a tradition of geographical scholarship and there is plenty of scope for further elaboration.

Box 1.1 Key anarchist thinkers

	Dates	Place of birth	Key writings
William Godwin	1756–1836	Wisbech, UK	*An Enquiry Concerning Political Justice & its Influence on General Virtue & Happiness* (1793)
Max Stirner	1806–1856	Bavaria, Germany	*The Ego and His Own* (1845)

Box 1.1 *continued*

Pierre-Joseph Proudhon	1809–1865	Besançon, France	*What is Property?* (1840); *On the Creation of Order in Humanity* (1843); *The Philosophy of Poverty* (1846); *The General Idea of the Revolution in the Nineteenth Century* (1851); *Justice in the Revolution and the Church* (1858); *The Federal Principle* (1863)
Michael Bakunin	1814–1876	Tver, Russia	*Principles and Organization of the International Brotherhood* (1866); *God and the State* (1882)
Leo Tolstoy	1828–1910	Tula, Russia	*War and Peace* (1863–69); *Anna Karenina* (1874–82); *A Confession* (1882); *Resurrection* (1897–98); *What is Art?* (1897-98); *Patriotism and Government* (1900); *The Significance of the Russian Revolution* (1906)
Elisée Réclus	1830–1905	Ste-Foy-la-Grande, France	*La Nouvelle Géographie Universelle* (1878–94) (19 volumes); *Evolution et Révolution* (1898) *L'Homme et la Terre* (1905–08);
Peter Kropotkin	1842–1921	Moscow, Russia	*In Russian and French Prisons* (1887); *The Conquest of Bread* (1892); *Fields, Factories and Workshops Tomorrow* (1899); *Modern Science and Anarchism* (1901); *Mutual Aid* (1902); *The Great French Revolution 1789–1793* (1909)
Errico Malatesta	1853–1932	Casterta, Italy	*Fra Contadini* (*Between Peasants*) (1884); *Anarchy* (1891)
Emma Goldman	1869–1940	Russia	*Anarchism and Other Essays* (1910); *My Disillusionment in Russia* (1923); *My Further Disillusionment in Russia* (1924); *Living My Life* (1931)

Sources: Marshall, 1993; Miller, 1984

Box 1.2 The life of Peter Kropotkin

Prince Peter Alexeivich Kropotkin was born into the Russian aristocracy in
1842. His father was a high-ranking officer in the army, owning property in
Moscow and an estate with 12,000 serfs in Kaluga. As was typical of his class,
the young Kropotkin attended the military academy called the Corps of Pages
from his early teens and he actually served as a Page de Chambre to the new
Tsar, Alexander II. It was clear, however, that Peter was growing tired of this
environment, and was developing more radical ideas, for in his early twenties he
chose a posting with the Cossacks of the Amur in Siberia rather than opting for
a safer career. The five years spent in Siberia proved to be a turning point for the
developing revolutionary as he encountered a wild and uncharted landscape
alongside anarchist ideas amongst the exiles confined to the region. His
expeditions in the area proved to be the foundation of his later reputation as a
physical geographer, and in particular, Kropotkin developed a new theory about
the glaciology and the orography (the layout and alignment of the mountain
ranges) of Asia. Moreover, his contact with people who lived without state
control and regulation, building their own communities in such harsh
conditions, helped to cement his anarchism. As he wrote in his *Memoirs*:
'I lost in Siberia whatever faith in state discipline I had cherished before.
I was prepared to become an anarchist' (1962: 148).

Such interests were further stimulated when Kropotkin visited the Swiss
Jura in 1872. The watchmakers of the region were famous for their political
ideas and their communitarian lifestyles, and they had an enormous influence
on Kropotkin's developing anarchism. In addition, this visit to Western
Europe brought Kropotkin into contact with the First International and the
libertarianism of Michael Bakunin. On his return to Russia, Kropotkin sought
out like-minded souls in his homeland, joining the Chaikovsky Circle for two
years and sympathising with the peasant-based movement of the Narodniks.
As a result of such activity Kropotkin was arrested and imprisoned for the first
time in March 1874. Imprisoned in the notorious Peter and Paul Fortress in St
Petersburg, he was only able to escape after three years. Exiled, he then moved
back to Western Europe where he made new contacts in the UK, Spain, Italy and
Switzerland, helping to set up a new anarchist journal called *Le Révolte*.
Following his expulsion from Switzerland, Kropotkin was arrested in Lyon in
1882 where he was confined in prison until 1886. The French authorities were
petitioned for his release by 15 British professors, the Royal Geographical
Society, William Morris and Patrick Geddes, reflecting his international
reputation as a scholar and political thinker.

When he was 44, Kropotkin moved to London, where he was to live for
another 41 years. Here he was involved in the journal *Freedom*, gave regular
lectures across the country and continued to travel abroad. Kropotkin kept up
with his writing, although he led an increasingly quiet life – particularly when
his support for the First World War alienated him from others in the anarchist
movement. For the last three years of his life Kropotkin returned to Russia.
The excitement of revolution was soured by his fears over Bolshevik tactics,
however, and he died in February 1921 in a village outside Moscow. Over
100,000 attended the funeral of this anarchist thinker and geographer.

Sources: Kropotkin, [1899] 1962; Miller, 1976; Brietbart, 1981; Cook, 1990; Marshall, 1993

Box 1.3 Elisée Réclus and the Paris Commune

Elisée Réclus was born to a religious family (his father was a Protestant Pastor), in a small village in the Dordogne, France, in 1830. Reflecting the profession and interests of his father, he attended Berlin University to study Theology in 1851, although while he was there he attended some of the popular geography lectures delivered by Carl Ritter. This geographical interest was then further fuelled by travels to America and Ireland – where he witnessed the terrible devastation of famine – an experience that fed his growing interest in the socialist movement (see Chapter 5 for more on the Irish Famine). Thus it was that when Réclus returned to Paris in 1857 he had become a geographer and a radical, playing a key role in the Paris Geographical Society, in Bakunin's secret *Brotherhood* and in the First International.

It was not, however, until the dramatic events of the Paris Commune that Elisée and his brother Elie became more clearly identified with the libertarian, or anarchist, wing of the socialist movement. The Commune began on 18 March 1871, when the workers of Paris took over the government of the city, in revolt against the authoritarianism and hardships they associated with the practices of the Second Empire. The Commune allowed a new social order to bloom, as men and women took on new roles and defended the city against the forces of the French army. This island of urban liberty was a reality for 73 days, reinforcing the strength of those who proselytised for social revolution, and giving Réclus the opportunity to test his ideas out in practice.

In the street battles that ended the Commune, however, 25,000 men and women were killed and Réclus, like many others, was imprisoned and then exiled to Switzerland. There, he began to write geography books and travel guides alongside anarchist pamphlets, cementing his role in the international movement. Between 1876 and 1894 he published the 19-volume *La Nouvelle Géographie Universelle* (New Universal Geography), and between 1905 and 1908 the smaller, 6-volume, *L'Homme et la Terre* (Man and Earth). These detailed, comprehensive, geography texts sought to integrate different sources of information about each part of the globe, and politically they were designed to show how the world's resources could be distributed to improve social well-being. Moreover, by challenging those of his profession who colluded with the imperialist carve-up of what is now the developing world, Réclus sought to use geography as a means to improve understanding, and empathy, across borders – eroding the power of the imperialist state by fostering a universal humanitarian spirit between the peoples of each nation and territory. In language which echoes the environmental concerns of our age, Réclus looked at the ways in which people could live in harmony with each other, and in a sustained relationship with the natural world (which he referred to as equilibrium). This holistic approach was later sidelined by other approaches to regional geography, but the themes of his work remain remarkably resonant in the contemporary world.

Réclus moved to Brussels for the last 11 years of his life where he took part in founding the New University, establishing a Geographical Institute there in 1898. Here, he did some unpaid tutoring and lecturing work, continuing with his research and writing from which he supported his family. He died in 1905.

Sources: Dunbar, 1978, 1981; Fleming, 1988

The key tenets of anarchist thought

> It would be misleading to offer a neat definition of anarchism, since by its very nature it is anti-dogmatic. It does not offer a fixed body of doctrine based on one particular world-view. It is a complex and subtle philosophy, embracing many different currents of thought and strategy. Indeed, anarchism is like a river with many currents and eddies, constantly changing and being refreshed by new surges but always moving towards the wide ocean of freedom. (Marshall, 1993: 3)

At the risk of funnelling the currents and eddies of anarchism into too narrow a channel, anarchists can be characterised by their opposition to all authority and their desire for a new social order. Authority, as embodied in institutions such as the church, state, army, factory and family, is argued to restrict human creativity and development, while upholding select social interests. Anarchists have sought to dispense with all such centralised and hierarchical power and have proposed living in small-scale, self-governing communities where decision making is shared (for a good introduction, see Harper, 1987). In this brief introduction to anarchist ideas we look at each part of this equation in turn.

Anti-authoritarianism

> When we ask for the abolition of the State and its organs we are always told that we dream of a society composed of men better than they are in reality. But no; a thousand times, no. All we ask is that men should not be made worse than they are by such institutions! (Kropotkin, from *Anarchism: Its philosophy and ideal*, 1970: 134)

Anarchists believe that centralised, hierarchical institutions play an enormous role in shaping the way people think and behave. By centralising decision making and taking control away from ordinary people, such institutions are argued to stifle the ability of people to think and act for themselves (so, for example, the officers of the local and national state are appointed or elected to take on responsibility for planning, development and environmental protection for you – taking away local control). Indeed, for writers such as Pierre-Joseph Proudhon and Oscar Wilde, society can only advance when people feel able to question authority and tradition, making their own decisions and taking their own course through life:

> The more ignorant man is, the more obedient he is, and the more absolute confidence in his guide … At the moment that man inquires into the motives which govern the will of his sovereign, – at that moment man revolts. If he obeys no longer because the king commands, but because the king demonstrates the wisdom of his commands, it may be said that henceforth he will recognise no authority, and that he has become his own king. (Proudhon, from *Property is Theft* [1840], quoted in Woodcock, 1977: 65)

Disobedience, in the eyes of anyone who has read history, is man's original virtue. It is through disobedience that progress has been made, through disobedience and through rebellion. (Wilde, from *The Soul of Man Under Socialism* [1891], quoted in Woodcock, 1977: 72; see Chapter 4 for more on Oscar Wilde)

Anarchists suggest that a hierarchical society in which some people have power and authority over others is rather primitive, restricting the scope of the mental and creative activity of its subjects or citizens. Indeed, many anarchists have argued that such social hierarchy and differentials of power interfere with the 'natural social order' of human society in which people would choose to freely interact in creative co-operation with one another. In his theory of *mutual aid*, for example, Peter Kropotkin drew upon scientific research to suggest that animals, including humans, are naturally co-operative in the interests of self-preservation, and that without the influence of law, power and property, humans would form co-operative, sustainable, communities in which all could take part (see Box 1.4 for a summary of this thesis). Using his own interpretation of evolutionary science, Kropotkin was able to argue that humans are *naturally* social, co-operative and moral beings, as he explained:

We are not afraid to say 'Do what you will; act as you will'; because we are persuaded that the great majority of mankind, in proportion to their degree of enlightenment, and the completeness with which they free themselves from existing fetters, *will behave and act always in a direction useful to society*; just as we are persuaded beforehand that a child will one day walk on its two feet, and not all fours, simply because it is born of parents belonging to the genus *homo* (Kropotkin, from *Anarchist Morality* [1892], quoted in Kropotkin, [1902] 1987: 10; emphasis added)

Box 1.4 Kropotkin's theory of Mutual Aid

Kropotkin's *Mutual Aid* was published in 1902 as a counter to the social Darwinists who argued that competition is the cornerstone of human nature and that capitalism and individualism are an inevitable manifestation of the biological drive to survive. Drawing upon his observations of animal activity and human societies in Eastern Siberia and Northern Manchuria, Kropotkin countered that the survival of any species depended upon co-operation, rather than competition. By supporting other members of their communities, animals would be better able to meet the challenges of their environment and the battle for scarce resources, as he explained:

Those animals which acquire habits of mutual aid are undoubtedly the fittest. They have more chances to survive, and they attain, in their respective classes, the highest development of intelligence and bodily organisation. (*Mutual Aid*, [1902] 1987: 24)

Box 1.4 *continued*

By living in social groups or 'societies', Kropotkin suggested that animals, such as ants, termites, bees and birds, are better able to survive by pooling their capacity for life:

> Life in societies enables the feeblest insects, the feeblest birds, and the feeblest mammals to resist, or to protect themselves from, the most terrible birds and beasts of prey; it permits longevity; it enables the species to raise its progeny with the least waste of energy and to maintain its members ... it enables the gregarious animals to migrate in search of new abodes. (*Mutual Aid*, [1902] 1987: 60)

Moreover, Kropotkin suggested that further up the food chain, sociability becomes more reasoned and less instinctive, culminating in humans, who have been able to develop language and culture through collectivity and mutual support. Mutuality and community helps a species survive, and this, rather than competition, is argued to be the key feature of human existence.

For Kropotkin, this human sociability and community could be witnessed in the management of communal pastures in Switzerland, in the shared communal traditions of those in the developing world, and in the friendly societies and trade unions found in Britain. Yet by the nineteenth century, it was clear that other processes were also at work and Kropotkin needed to explain the existence of exploitation, war and oppression among people said to be 'naturally moral' and 'biologically disposed' to mutual concern. Kropotkin suggested that the 'natural order' of human life was being erased by the influence of the state and other authorial institutions of social control, as he explained:

> The absorption of all social functions by the state necessarily favoured the development of an unbridled narrow-minded individualism. In proportion as the obligations towards the state grew in numbers the citizens were evidently relieved from their obligations towards each other. (*Mutual Aid*, [1902] 1987: 183)

Kropotkin was thus able to blame the state and its attendant power relations for distorting 'natural' human society and eroding the co-operation and mutuality which would otherwise need to exist. And as an anarchist, Kropotkin pointed to examples of solidarity and collectivity as evidence of our 'true' human nature which could reassert itself in an alternative world:

> In short, neither the crushing powers of the centralized state nor the teachings of mutual hatred and pitiless struggle which came, adorned with the attributes of science, from obliging philosophers and sociologists, could weed out the feeling of human solidarity, deeply lodged in men's understanding and heart, because it has been nurtured by all our preceding evolution. (*Mutual Aid*, [1902] 1987: 229)

The theory of mutual aid was designed to provide a scientific foundation for anarchism, and it has been important in presenting anarchism as more than a naive and idealistic set of ideas. By arguing that the end of the state would allow humans to live as nature intended, in harmony with each other, Kropotkin sought to demonstrate that anarchism had strong biological roots.

Predicated on the view that it is human nature to co-operate, anarchists have thus targeted those institutions which are argued to impose hierarchies of power and control over the 'natural' order of society. In particular, the state and government have been condemned as socially repressive, enforcing laws which stifle individual decision making and action, while also upholding the entrenched interests of those who have power and wealth. Laws to protect private property, to restrict unionisation and to control political organisation are all seen as evidence that the state and its government act to defend existing inequalities, ensuring that the rich and powerful are protected. Rather than act to redistribute wealth, eradicate poverty and improve the living standards of the majority, anarchists argue that governments always end up protecting the wealthy. And this, they suggest, is due to the damaging impact of hierarchical social organisation, whereby those with authority and power will always act, and put pressure on others to act, to defend their privilege and control. In colourful prose, Proudhon articulated this critique of the state and government in his book, *The General Idea of the Revolution in the Nineteenth Century*:

> To be GOVERNED is to be at every operation, at every transaction, noted, registered, enrolled, taxed, stamped, measured, numbered, assessed, licensed, authorized, admonished, forbidden, reformed, corrected, punished. It is, under pretext of public utility, and in the name of general interest, to be placed under contribution, trained, ransomed, exploited, monopolized, extorted, squeezed, mystified, robbed; then, at the slightest resistance, the first word of complaint, to be repressed, fined, despised, harassed, tracked, abused, clubbed, disarmed, choked, imprisoned, judged, condemned, shot, deported, sacrificed, sold, betrayed; and, to crown all, mocked, ridiculed, outraged, dishonoured. That is government; that is its justice, that is its morality. (Proudhon [1851], quoted in Miller, 1984: 6)

And in language which reflected the socialist arguments of his day, Errico Malatesta highlighted the powerful interests behind government legislation and its enforcement:

> The basic function of government everywhere at all times, whatever title it adopts and whatever its origins and organization may be, is always that of oppressing and exploiting the masses, of defending the oppressors and exploiters. (Malatesta, from *Anarchy*, [1891] 1974: 20–21)

This critique of the state and government – and the authorial relations represented – is common to all anarchists, even those who are prepared to accept the necessity of other laws or guidelines to shape social affairs. The pacifist anarchist Leo Tolstoy, for example, condemned the laws of government while also proselytising the moral laws distilled in his reading of Christianity:

> The truth is that the state is a conspiracy designed not only to exploit, but above all to corrupt its citizens ... I understand moral laws, and the laws of morality and religion, which are not binding, but which lead people forward

and promise a harmonious future; and I sense the laws of art which always bring happiness; but the laws of politics are such terrible lies for me … and I will never serve *any* government anywhere. (Tolstoy, quoted in Marshall, 1993: 364; emphasis in the original)

Likewise, Michael Bakunin was prepared to recognise the authority associated with particular skills and knowledge as long as such authority was not based on power and control over others, as he explained in his volume *God and the State*:

> Does it follow that I reject all authority? Far from me such a thought. In the matter of boots, I refer to the bootmaker; concerning houses, canals or railroads, I consult that of the architect or the engineer … But I allow neither the bootmaker not the architect … to impose his authority upon me. I listen to them freely and with all the respect merited by their intelligence, their character, their knowledge, reserving always my incontestable right of criticism and censure … I recognise no infallible authority, even in special questions; consequently, whatever respect I may have for the honesty and the sincerity of such and such an individual, I have no absolute faith in any person. Such a faith would be fatal to my reason, to my liberty, and even to the success of my undertakings; it would immediately transform me into a stupid slave, an instrument of the will and interests of others. (Bakunin, from *God and the State* [1883], quoted in Woodcock, 1977: 313)

Echoing the mutualist arguments of Proudhon and Kropotkin, both Tolstoy and Bakunin distinguish between the social order which is said to come *from within*, guiding our decision making and our personal conduct, and that imposed *from without*. Anarchists have acknowledged that the social order generated by individual action can only be founded upon relationships made locally through face-to-face contact.

Creating a new social order

> It [communist society] cannot exist without creating a continual contact between all for the thousands and thousands of common transactions; it cannot exist without creating local life, independent in the smallest unities – the block of houses, the street, the district, the commune. It would not answer its purpose if it did not cover society with a network of thousands of associations to satisfy its thousand needs: the necessaries of life, articles of luxury, of study, of enjoyment, amusements. And such associations cannot remain narrow and local; they must necessarily tend (as is already the case with learned societies, cyclist clubs, humanitarian societies and the like) to become international. (Kropotkin, from *Anarchism: Its philosophy and ideal*, 1970: 140)

In a society without any centralised control and regulation emanating from the state or government, social order can only be forged through the co-ordinated decision making of individuals. Such co-ordination, however, requires considerable face-to-face contact to ensure that the actions of each person do not negatively impact on the interests of others. And for this reason, an anarchist society could only take shape at a very small scale. By forging small-scale, decentralised communities where individuals know each other, share common interests and act upon collective decisions, anarchists argue that a new world can be made. Moreover, it is suggested that the localisation of power and decision making would prompt individuals to consider the needs of their environment as well as the people around them. By sustaining the needs of each other, communities would also have to sustain the land and production on which they survived.

In contrast to the experience of work in a capitalist world, in which divisions of labour and power ensure that individuals specialise in particular tasks, often for very long periods of their lives, anarchists have argued that the small-scale community could foster a new pattern of work (for further elaboration of the anarchist critique of work in the capitalist system, see Box 1.5). In particular, anarchists have sought to unite mental and manual work in the labours of each commune member, as Kropotkin outlined in *Fields, Factories and Workshops Tomorrow*:

> Political economy has hitherto insisted chiefly upon division. We proclaim integration; and we maintain that the ideal of society – that is, the state towards which society is already marching – is a society of integrated, combined labour. A society where each individual is a producer of both manual and intellectual work; where each able-bodied human being is a worker, and where each worker works both in the field and in the industrial workshop; where every aggregation of individuals, large enough to dispose of a certain variety of natural resources – it may be a nation or rather a region – produces and itself consumes most of its agricultural and manufactured produce. (Kropotkin, [1899] 1985: 26)

Moreover, as is suggested in the final part of this quotation, a decentralised anarchist society would allow communities to be more or less self-sufficient, consuming their resources and products at a localised scale. Through federation, each community could be connected to others, but they would exchange only essential goods to protect the environment and local resources for the long term.

Box 1.5 The anarchist critique of work in the capitalist system

The labour process in the capitalist system ensures that workers work to produce profit for their employer while earning wages to live. This process means that individuals are not in control of their labour, and that many of the pleasures of work are destroyed. For the anarchist, the pleasures of tilling the land or crafting something from wood are fundamental to being human, and it is such labour of love that is at the heart of their vision for an alternative world. As the following quotations attest, anarchists have been outspoken critics of the capitalist labour process and they seek to replace it with a holistic approach to employment:

Box 1.5 *continued*

> Under present conditions there is little choice given the average man to devote himself to the tasks that appeal to his leanings and preferences. The accident of your birth and social station generally predetermines your trade or profession … Is it any wonder, then, that most people, the overwhelming majority, in fact, are misplaced? … Necessity and material advantages, or the hope of them, keep most people in the wrong place … The things the craftsmen produced in the days before modern capitalism were objects of joy and beauty, because the artisan loved his work. Can you expect the modern drudge in the modern factory to make beautiful things? He is part of the machine, a cog in the soulless industry, his labour mechanical, forced. Add to this his feeling that he is not working for himself but for the benefit of someone else, and that he hates his job or at best has no interest in it except that it secures his weekly wage … Under anarchism each will have the opportunity for following whatever occupation will appeal to his natural inclinations and aptitude. Work will become a pleasure instead of the deadening drudgery it is today. (Berkman, *What is Anarchist Communism?* [1929], in Woodcock, 1977: 335–337)
>
> In the present state of society, when we see Cabinet Ministers paying themselves thousands a year, whilst the working man has to content himself with less than a hundred; when we see the foreman paid twice or three times as much as the ordinary hand, and when amongst workers themselves there are all sorts of gradations … we say, 'Let us have done with privileges of education as well as of birth.' We are Anarchists just because such privileges disgust us. (Kropotkin, *The Wage System* [1889], in Woodcock, 1977: 354)

There has been a long tradition of anarchist sabotage in the workplace, illustrated by the character Souvarine in Zola's novel *Germinal*, who blows up a mine shaft in response to the employers' refusal to meet strikers' demands. In addition, however, anarchists have developed their most permanent forms of organisation amongst workers by linking the struggle for better wages and conditions with the wider battle for a new social order. This form of anarchism is called *anarcho-syndicalism* and has been particularly powerful in France and Spain. As George Woodcock summarises in the following quotation, anarchist unions have a different organisational form than that practised by their more conventional counterparts:

> The syndicate is a form of union which differs from the ordinary trade union in that it aims, not only at the gaining of improvements in wages and conditions under the present system, but also at the overthrow of that system by a social revolution based on the economic direct action of the workers. This is not to say that it ignores the day-to-day struggle, but its members recognize that only by a complete destruction of the structure of property and authority can justice and security ever be attained by the workers.
>
> The syndicate differs also from the ordinary trade union in its method of organization. The ordinary trade union follows the pattern of governmental society in that it has a centralized form, with authority at the centre and a permanent bureaucracy who, like any other bureaucracy, rapidly gain privilege and power and rise into a class with an economic position considerably higher than that of the workers who pay them and whom they are supposed to serve. The syndicate, on the other hand, is based on the organization of the workers

Box 1.5 *continued*

by industry at the place of work. The workers of each factory or depot or farm are an autonomous unit, who govern their own affairs and who make all the decisions as to the work they will do. These units are joined federally in a syndicate which serves to coordinate the actions of workers in each industry. The federal organization has no authority over the workers in any branch, and cannot impose a veto on action like a trade union executive. It has no permanent bureaucracy, and the few voluntary officials are chosen on a short term basis, have no privileges which raise their standard of living above that of the workers, and wield no authority of any kind. (Woodcock, *Railways and Society* [1943], in Woodcock, 1977: 209–210)

This form of trade unionism, which localises decision making and authority, took root in parts of Spain and France during the late nineteenth and early twentieth centuries. The traditions of direct action, associated with syndicalism, are still evident in French union protests and Spanish union organisation. Moreover, the institutional strength of the syndicates, compared to other forms of anarchist organisation, has allowed anarcho-syndicalists to co-ordinate internationally in the *International Working Men's Association* (Woodcock, 1962). This body was founded in Berlin, in 1922, when anarchist trade unionists became disillusioned with the Bolshevik leadership of the Red International. At its height, this body had three million members, concentrated in France, Spain and Italy, linking trade union activity across space (for more on labour internationalism, see Chapter 2).

As we will see later on in this chapter, some anarchists have sought to establish such new communities in the here-and-now before any widespread revolution in social relations takes place. And so too, many anarchists have sought to intervene in the organisation and experience of the capitalist labour process, arguing that it is both alienating for, and exploitative of, those doing the work. Following the 1848 revolutions, for example, Proudhon set up a People's Bank which valued commodities on the basis of the labour time spent in production. Reflecting the input made by the worker, rather than any market price or the profit margins of the employer, the bank was able to value, buy and sell products using labour time, rather than profit, as the unit of cost. In less than a year the People's Bank attracted 27,000 members in Paris, drawing workers into local mutual associations where they exchanged commodities with each other at the prices set by the bank (see Woodcock, 1956). Likewise, many anarchists have been involved in setting up co-operative businesses within the confines of the capitalist system as a means to share tasks, distribute wealth and act to promote an alternative social order. As Tolstoy declared in his bid to foster a new world in the present:

> [T]he founding of cooperatives and participation in them is the only
> social activity which a moral, self-respecting person who doesn't wish to
> be a party to violence can take part in our time. (Tolstoy, quoted in Marshall,
> 1993: 378)

Anarchists are thus both reformists and revolutionaries, acting in the short term as well as dreaming of a new society in the post-revolutionary future. In addition to their involvement in co-operatives, communities and communes (where groups of like-minded people live together outside the confines of the traditional nuclear family), anarchists have also taken a keen interest in educational reform. Tolstoy established a school for the peasant children on his estate in 1859, in which he encouraged the children to learn through doing and thinking for themselves. Similarly, the feminist anarchist Emma Goldman involved herself in the Modern School Movement in the United States, while Elisée Réclus urged his geography students at the Institute for Advanced Studies in Brussels to question him and to take part in seminars and informal exchange (Fleming, 1988: 193). Focusing on the individual, rather than any mass political movement, the anarchist thus seeks long-term social change through practical experiment and example in the short term, as Andrew Rigby outlines:

> [T]he anarchist revolution is not so much an *event* as a process, a *process* of undermining all existing institutions and relationships: how we live, how we learn, how we dress, how we think. Everything. As such, living an alternative lifestyle is not just about having a good time, but trying to create 'the good life' in the here-and-now – exercising choice, claiming autonomy, practising mutual aid – as part of the process of undermining the state and its related institutions and practices. (Rigby, 1990: 54; emphasis in the original)

Anarchism, in all its variety, focuses on the individual as the mechanism of social upheaval and as the means to build a new world, and it is this emphasis upon individual *will* which most clearly distinguishes anarchism from the marxist tradition, as George Woodcock explains:

> For [anarchists] ... history does not move, as it does for the marxist, along the steel lines of dialectical necessity. It emerges out of struggle, and human struggle is a product of the exercise of man's will, based on the spark of free consciousness within him, responding to whatever impulse – in reason or in nature – provokes the perennial urge to freedom. (Woodcock, 1962: 27)

While nineteenth-century marxists argued that workers, led by a vanguard revolutionary party, should seize the state to implement change, anarchists were advocating non-hierarchical organisation to dismember the state and all institutions of power (see Box 1.6 for a summary of the split between anarchism and marxism). Moreover, even though anarchists have always disagreed in their views about the need for violence in the transition to a new world, they have all tended to eschew permanent political organisation in favour of the individual taking part in direct action in response to a particular cause. Acts of mass civil disobedience, direct action, urban unrest, industrial sabotage or political murder have all been part of anarchist strategies to secure a new society, and each depends upon the individual rather than a wider party or group. Only anarchist organisation amongst workers, known as *anarcho-syndicalism*, retains a more permanent structure, based on collective militancy in the workplace (see Box 1.5).

Box 1.6 The split between anarchists and marxists in the International Working Men's Association

The International Working Men's Association (known as the First International) was founded in 1864 uniting socialists of many persuasions to spread the class struggle across national divides (see also Chapter 2). In the early years of this new organisation, marxists organised alongside reformist trade unionists and libertarian, or anarchist, socialists, seeking to foster solidarity between workers in one country with those in another. However, relationships were often hostile, and it soon became clear that Marx and his followers had a very different model of revolutionary socialist organisation from their more libertarian comrades. Marx published attacks on the mutualist ideas of Proudhon (see *The poverty of philosophy*, [1847] 1950) and he later came head to head with the anarchism of Michael Bakunin. Accused of co-ordinating a separate group of revolutionaries within the International, Bakunin was expelled, in his absence, at the 1872 Congress, held in the Hague.

In arguments which now appear remarkably prescient in the light of events which have unfolded in Russia after the revolutions of 1917, the anarchist critique of Marx and his followers centred on the risks associated with authoritarian political organisation and on the dangers of seizing the capitalist state as a means to implement social change. For Bakunin, freedom was a necessary component of socialist organisation and social change, and it could not be compromised by the requirements of the revolution, as he explained:

> Equality without freedom is the despotism of the state ... the most fatal combination that could possibly be formed, would be to unite socialism to absolutism; to unite the aspiration of the people for material well-being ... with the dictatorship or the concentration of all political and social power in the state ... We must seek full economic and social justice only by way of freedom. There can be nothing living or human outside of liberty, and a socialism that does not accept freedom as its only creative principle ... will inevitably ... lead to slavery and brutality. (Bakunin, quoted in Dolgoff, 1980: 4)

The division between marxist and anarchist socialists remains in place to this day, and as the following quotation from Kropotkin attests, the two traditions have developed very different organisational styles; the marxists favouring a disciplined party which can seize control of the organs of social power (as was demonstrated in the Russian Revolution of 1917), and the anarchists adopting very loose, non-hierarchical associations for the pursuit of particular goals:

> Now it is the workers' and peasants' initiative that all parties – the socialist authoritarian party included – have always stifled, wittingly or not, by party discipline. Committees, centres, ordering everything; local organs having but to obey, 'so as not to put the unity of the organization in danger'. (Kropotkin, *Anarchism: Its philosophy and ideal*, 1970: 142)

In the light of the revolutionary events of the twentieth century, anarchists have been vindicated in their warnings about centralised leadership and the power of a socialist state. As Marshall writes in his summary of the two traditions of socialist thought:

Box 1.6 *continued*

> While Marx may have won the initial dispute within the International, subsequent events have tended to prove the validity of Bakunin's warnings about centralism, state socialism, and the dictatorship of the proletariat. He had prophetic insight into the nature of communist states which have all become to varying degrees centralised, bureaucratic and militaristic, ruled by a largely self-appointed and self-reproducing élite ... Bakunin, not Marx, has been vindicated by the verdict of history. (Marshall, 1993: 305)

The most celebrated theoretician of anarchist practice has been the Russian, Michael Bakunin. In 1866, he laid down the *Principles and Organization of the International Brotherhood*, a framework for the secret network of anarchist revolutionaries that he recruited to propagate the anarchist message by word and deed. Designed to agitate for social revolution, *The Brotherhood* was also structured to avoid the dangers of authoritarianism and control that Bakunin associated with Marx, as he explained:

> Our aim is the creation of a powerful but always invisible revolutionary association which will prepare and direct the revolution. But never, even during open revolution, will the association as a whole, or any of its members, take any kind of public office, for it has no aim other than to destroy all government and to make government impossible everywhere ... It will keep watch so that authorities, governments and states can never be built again. (Bakunin, quoted in Dolgoff, 1980: 10)

In addition, Bakunin, in common with other anarchists, sought influence amongst a wide range of social groups beyond the working class, involving peasants, racial minorities and the unemployed, in his quest for a new social order. All were welcome to the ranks of the anarchist tradition, as long as they refused to take control over the decisions of others:

> Hierarchical order and promotion do not exist, so that the commander of yesterday can become a subordinate tomorrow. No one rises above the other, or if one does rise, it is only to fall back again a moment later, like the waves of the sea returning to the salutary level of equality. (Bakunin, quoted in Dolgoff, 1980: 12)

Bakunin's *Brotherhood* remained a rather shady organisation, and little is known of its exact size or influence, although Bakunin himself took part in many revolutionary adventures during his life (for a brief summary, see Box 1.7).

Box 1.7 The revolutionary activity of Michael Bakunin

Bakunin's earliest experience of rebellion and resistance came when he was serving in the Russian army in Poland at the time of the 1832 insurrection. This experience caused him to leave the army and launch himself on a revolutionary career which took him all over Europe. He was in Germany and Prague during the revolutions of 1848 and he took a major part in a popular uprising in Dresden in 1849. Following periods in prison in Germany, Austria and Russia, Bakunin escaped from the Peter and Paul Fortress in St Petersburg in 1857 and then travelled extensively in Europe, building up his revolutionary associations and contacts. By 1870 he was in Lyon, France, taking part in an uprising which culminated in the seizure of the town hall, after which the following poster was issued, reflecting Bakunin's libertarian dream:

ARTICLE 1: The administrative and governmental machinery of the state, having become impotent, is abolished.

ARTICLE 2: All criminal and civil courts are hereby suspended and replaced by the People's justice.

ARTICLE 3: Payment of taxes and mortgages is suspended. Taxes are to be replaced by contributions that the federated communes will have collected by levies upon the wealthy classes, according to what is needed for the salvation of France.

ARTICLE 4: Since the State has been abolished, it can no longer intervene to secure the payment of private debts.

ARTICLE 5: All existing municipal administrative bodies are hereby abolished. They will be replaced in each commune by committees for the salvation of France. All governmental powers will be exercised by these committees under the direct supervision of the People.

ARTICLE 6: The committee in the principal town of each of the nation's departments will send two delegates to a revolutionary convention for the salvation of France.

ARTICLE 7: This convention will meet immediately at the town hall of Lyon, since it is the second city of France and the best able to deal energetically with the country's defence. Since it will be supported by the People this convention will save France.

TO ARMS!!

(From Marshall, 1993: 286)

He died, internationally famous for his anarchist deeds and revolutionary fervour, in Switzerland during July 1876.

Sources: Guillaume, 1980; Marshall, 1993

In contrast to Bakunin's advocacy of secret revolutionary organisation and the violent overthrow of authorial social order, other anarchists have advocated non-violent direct action as the means to achieve a new world. Leo Tolstoy, for example, is known to have influenced Mahatma Gandhi's nationwide campaign of mass civil disobedience for Indian independence in the years before and after the Second World War (also see Chapter 5). And likewise, Emma Goldman's plea for women's liberation through personal political action anticipated the agenda of the women's movement in the western world from the 1960s to the present day (see Box 1.8 and Chapter 3). It is this less violent strain of anarchism that is still influential in radical politics, underlying movements for peace, for political freedoms and environmental protection. Moreover, these more practical aspects of anarchist thought and deed have left their mark on the contemporary landscape.

Box 1.8 The anarcho-feminism of Emma Goldman

Emma Goldman was an anarchist and a feminist writer and activist. Born in Russia, she spent most of her life in the United States and in Britain. In contrast to other feminists of the day, who supported the campaign for suffrage, she anticipated many of the claims of the later women's movement by focusing on the individual freedom which had to come from within. As she declared in the following statement, women had to free themselves, rather than demanding that the State give them freedom by granting them a franchise equal to men:

> First, by asserting herself as a personality, and not as a sex commodity. Second, by refusing the right to anyone over her body; by refusing to bear children, unless she wants them; by refusing to be a servant to God, the State, society, the husband, the family etc., by making her life simpler, but deeper and richer ... Only that, and not the ballot, will set women free, will make her a force hitherto unknown in the world, a force for real love, for peace, for harmony; a force for divine fire, of life-giving; a creator of free men and women. (Goldman, *Anarchism and Other Essays*, [1910] 1969: 407)

Emma Goldman refused marriage, practised 'free love' (or open relationships in contemporary terms), advocated and disseminated the means of birth control, and supported the equality of the sexes. Her anarcho-feminism suggested that women could not be free unless they freed themselves, and freed society from the Church, State and the nuclear family. For the first time, she united the cause of anarchism with the battle for women's liberation and equality (see Chapter 3 for more on feminist politics).

The uneven geography of anarchism

As might be expected, anarchist ideas have taken root more deeply in some places than others. As Woodcock explains, the antecedents of anarchist thought were in England, its key philosophers have been Russian, the French have contributed in thought and deed, and the widest practical impact of the tradition has been felt in Spain:

> In England, with Winstanley and Godwin, anarchism first appeared as a recognizable social doctrine. In Spain it attained its largest numerical support.

> In Russia it produced, with Kropotkin, Bakunin and Tolstoy, its most distinguished group of theoreticians. Yet for many reasons it is France that deserves pride of place among the countries that have contributed to the anarchist tradition. (Woodcock, 1962: 257)

Such uneven geographical development is a product of both individuals and circumstances. Ideas need to be in currency but there also needs to be fertile ground in which such ideas can take root and grow over time. As far as Peter Marshall is concerned, for example, there are clear reasons why Russia produced so many of the intellectuals who played such an important part in nineteenth-century anarchism:

> It is hardly a coincidence that the Russian aristocracy should have produced three of the greatest anarchist thinkers in the nineteenth century in Bakunin, Kropotkin and Tolstoy. They were all able to witness at close quarters the tyranny of the Tsarist regime, and, conversely, the inspiring example of peasant communities living in an orderly and peaceful fashion without trace of government. (Marshall, 1993: 382)

Just as their experiences prompted them to develop anarchist ideas, so too, the circumstances of the people to whom these ideas were taken also determined the extent to which they made sense. In many cases, it was those groups of people who were most threatened by the political and economic centralisation of governmental power during the nineteenth century that turned to anarchist ideas. The individualism of libertarian thought appealed to groups of artisans, peasants and the poor who were losing their way of life. Such groups were concentrated in parts of France, Spain, the Swiss Jura, Italy and Russia, and it was here that anarchism took greatest hold:

> [T]he countries and regions where anarchism was strongest were those in which industry was least developed and in which the poor were poorest. As progress engulfed the classic fatherlands of anarchism, as the factory workers replaced the handcraftsmen, as the aristocrats became detached from the land … anarchism began to lose the main sources of its support. (Woodcock, 1962: 445)

It is not surprising that the division between marxists and anarchists which came to a head at the 1872 Congress of the International Working Men's Association had geographical manifestations (see Box 1.6). Marx found his strongest supporters in the British, German and Austrian sections of the International, while Bakunin had supporters in the Swiss Jura, Spain and Italy. Such divisions reflected the circumstances of people in each place during this time, giving differential purchase to different political ideas and agendas over space.

Moreover, after the split with the marxists in the First International, anarchists found it very difficult to co-ordinate at an international scale, and in this regard, local, regional and national differences in anarchist ideas, organisation and experiment tended to persist. A Black International, to match their Red socialist rival, was always threatened, but congresses were held only in London in 1881 and in Amsterdam in 1907. The latter meeting managed to agree to the following resolution, but it reflects the wide differences between the delegates who attended:

> The anarchists urge their comrades and all men aspiring to liberty, to struggle according to circumstances and their own temperaments, and by all means – individual revolt, isolated or collective refusal of service, passive and active disobedience and the military strike – for the radical destruction of the instruments of domination. They express the hope that all the peoples concerned will reply to any declaration of war by insurrection and consider that anarchists should give the example. (Quoted in Woodcock, 1962: 250)

In the event, this was the last successful attempt to organise at this scale, and only the anarcho-syndicalists have proved able to co-ordinate across borders in the long term. The anarchist movement has mainly relied on particular agitators to forge links between supporters in different locations, and it is significant that the key protagonists of the nineteenth century knew one another and often moved from one place to another, sometimes doing so to escape persecution:

> Anarchist literature passed freely from country to country, and the works of men like Bakunin, Kropotkin and Malatesta were translated into many languages. In addition to this exchange of ideas and propaganda there was also a constant intercourse between anarchist militants, owing largely to the fact that the life of the dedicated revolutionary often forced him to go into temporary exile or even to seek an entirely new home abroad. Errico Malatesta agitated and conspired not only in Italy, but also in France, England, Spain, the Levant, the United States and Argentina, and there were many like him ... Anarchism was international in theory and to a great extent in practice even if it was only sporadically so in organizational terms. (Woodcock, 1962: 248–249).

Thus anarchism has remained a diverse phenomenon, differentiated across space and time. Geographically, it has been more influential in some places than in others, and its most spectacular impact has been felt in Spain. The nineteenth-century roots of anarchist organisation in Spain were clearly manifest during the 1936–39 Civil War when activists had the opportunity to put their ideas into practice on a wide scale (see Figures 1.1 and 1.2). When the new Republican government was attacked by General Franco and his fascist-inspired military forces on 18 July 1936, large areas of Spain came under anarchist influence as the people organised to defend their freedom and human rights. Anarchism had been established in Spain during the nineteenth century, and the anarcho-syndicalist Confederación Nacional del Trabajo (CNT) and the anarcho-communists in the Federación Anarquista Ibérica (FAI) had considerable weight in Catalonia, the Levante, New Castile and Andalusia.

Figure 1.1 The Spanish Civil War: volunteers leave Madrid to fight the rebels

In these areas, rural peasants and urban workers created a social revolution to defend themselves against the might of the Francoist army. For the first time they were able to collectivise the land, take over the factories and produce for collective need rather than individual profit. Approximately 2000 rural collectives were formed on over 15 million acres of land that was simply expropriated from rich landlords who had previously neglected the needs of the people (Breitbart, 1978: 60; see Figure 1.3).[1] Anarchist principles were put into practice by allowing local control, federating collectives with each other and involving as many people as possible in the decisions determining everyday life:

> An administrative committee would be elected, but this would operate under the constant supervision of the population, meeting at least once a week in full assembly to hasten the achievement of free communism. In the factories the process was similar, with a workers' committee becoming responsible to the general assembly of the syndicate, and technicians (in a few cases the former owners or managers) planning production in accordance with the workers' views. (Woodcock, 1962: 371)

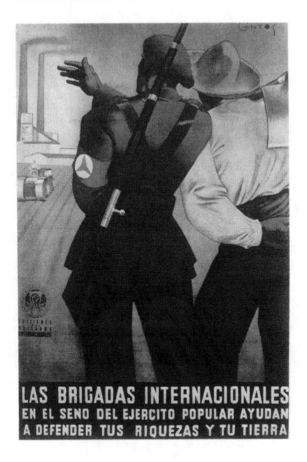

Figure 1.2 International solidarity in the Spanish Civil War

Between July and October 1936 Barcelona was almost completely under workers' control and the revolution extended to medicine, education, transportation and cultural affairs (see Fraser, [1979] 1994; Miller, 1984; Marshall, 1993; Amsden, 1978). George Orwell wrote *Homage to Catalonia* based on his experiences in Republican Spain, providing the following eye-witness account of life in Barcelona during 1936:

> It was the first time I had ever been in a town where the working class was in the saddle. Practically every building of any size had been seized by the workers and was draped with red flags or with the red and black flag of the anarchists ... Every shop and cafe had an inscription saying that it had been collectivised; even the bootblacks had been collectivised and their boxes painted red and black. Waiters and shop-workers looked you in the face and treated you as an equal. Servile and even ceremonial forms of speech had temporarily disappeared ... Tipping was forbidden by law; almost my first experience was receiving a lecture from a hotel manager for trying to tip a lift boy.

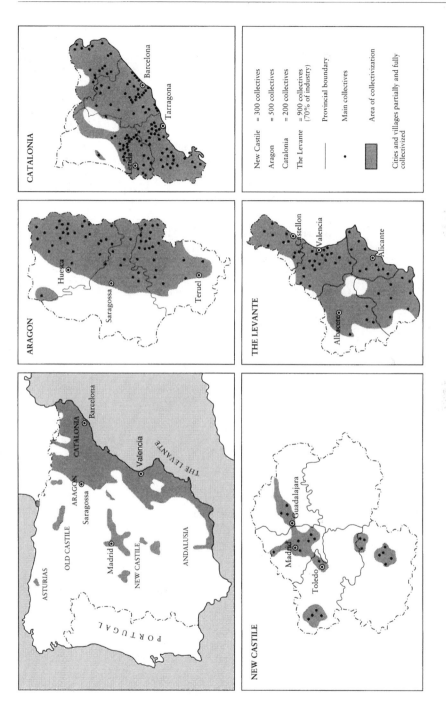

Figure 1.3 Major areas of collectivisation during the Spanish Civil War (*Source*: Breibart, 1978)

> There were no private motor cars, they had all been commandeered, and all the trams and taxis and much of the other transport was painted red and black. The revolutionary posters were everywhere, flaming from the walls in clean reds and blues that made the few remaining advertisements look like daubs of mud. Down the Ramblas, the wide central artery of the town where crowds of people streamed constantly to and fro, the loudspeakers were bellowing revolutionary songs all day and far into the night. And it was the aspect of the crowds that was the queerest thing of all. In outward appearance it was a town in which the wealthy classes had practically ceased to exist ...
>
> Above all, there was a belief in the revolution and the future, a feeling of having suddenly emerged into an era of equality and freedom. Human beings were trying to behave as human beings and not as cogs in the capitalist machine. (Orwell, [1938] 1989: 8–10)

Even after its eventual defeat in 1939, partly through divisions within Republican ranks, anarchism had reshaped the landscape of Spain, as Breitbart (1978; see also Garcia-Ramon, 1978) remarks:

> A revolution that began by altering social and economic relationships thus created totally new spatial formations. (Breitbart, 1978: 65)

The Spanish Civil War allowed anarchists to demonstrate the role that community, collectivisation and social equality could play in reshaping everyday life. Until then, anarchist experiments had been on a much smaller scale, usually involving small numbers of pioneers who sought to establish settlements in new locations, as discussed in the following section.

Redrawing the map: Anarchist settlements in the UK

As we have seen, anarchists argue that small-scale communities are the ideal setting for the practical expression of a new social order. During the Spanish Civil War, anarchists overturned the power relations and property rights in *existing* rural and urban communities, but in most places, anarchists have sought to establish *new* settlements as a means to put their ideas into practice. One of the earliest of such ventures on English soil was led by the Diggers, a movement of people on the left of the Republican movement, during the Civil War. Gerrard Winstanley led about 40 settlers onto common land at St George's Hill, Surrey, in April 1649 to 'work in righteousness and lay the foundation of making the earth a common treasury for all' (Marshall, 1993: 97; see also Armytage, 1961; Hardy, 1979, 1990). These pioneers sought to overturn the prevailing social order by establishing a new settlement and managing the land collectively, sharing tasks and providing for the needs of each community member. At the time, the language of religion was used to express political ideals, and the Diggers, Levellers and Ranters argued that God gave them the right to liberty and to govern themselves. Indeed, these dissenters believed they could make unmediated contact with God and the Holy Spirit, dispensing with priests, vicars and the church hierarchy (see Thompson, 1994; Marshall, 1993). Although the St George's Hill settlement lasted only a year, declining in the face of severe hostility, the ideas of these dissenters had much more lasting effect.

More than 200 years later, anarchist sentiment inspired a number of experimental settlements in different parts of the country (see Figure 1.4).[2] During the 1890s followers of Kropotkin founded the short-lived *Clousden Hill Free Communist and Co-operative Colony* at Forest Hall, Newcastle, and the *Norton Colony*, near Sheffield (Hardy, 1990: 39). In addition, the ideas of Tolstoy took particular hold amongst groups of free thinkers who sought to lead a simpler, more useful and communitarian life. Bruce Wallace founded the Brotherhood Trust in January 1894 with the aim of establishing a co-operative commonwealth to penetrate the capitalist system and undermine it

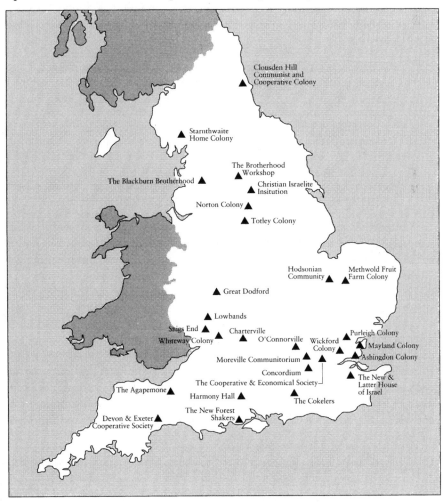

Figure 1.4 Alternative communities in nineteenth-century England (from Hardy, 1979: 15)

from within. In order to raise money to buy land to settle, the Brotherhood ran a grocery and vegetable co-operative at 1 and 5 Downland Road, Kingsland, North London. The Brotherhood churches spread in London and by 1897 the Croydon Brotherhood had sufficient resources to buy land at Purleigh in Essex. Started by five individuals, the community numbered more than 60 in only a year. In its wake, communities sprang up in other parts of Essex and further afield: the *Ashingdon Colony* (1897) and the *Wickford Colony* (1898) in Essex, the *Brotherhood Workshop* (1897) in Leeds, the *Blackburn Brotherhood* (1899) and the *Whiteway Colony* (1898) in the Cotswolds (Hardy, 1990: 40). The Whiteway Colony was formed by a handful of settlers who had grown disillusioned with life at the larger Purleigh Colony where disagreements raged over who should be able to join the community and the degree to which they should be involved in commercial life. Whiteway, in contrast, was to be more of a 'pure' application of anarchist ideals, as Dennis Hardy explains:

> Whiteway was to last longer than other communities, and was from the start a spirited attempt to put anarchist principles into practice. The colonists who arrived at the Cotswold site in August 1898 were steeped in Tolstoyan doctrine. They shared a rejection of urban life in favour of the country, they opposed any form of private property (some of them refusing to use money at all times), they were pacifist and practised non-resistance in their own lives, they worked towards equality for women in the community, they favoured free union rather than state marriage, they looked for more liberated forms of clothing for both men and women, and (like so many of their predecessors) their pacifist philosophy was matched with a vegetarian diet. (Hardy, 1990: 44)

Whiteway began when Daniel Thatcher, a journalist, and Joseph Burtt, who had worked in a bank, put together £450 to buy a large house and 40 acres of land near Stroud in Gloucestershire. Joined by seven other men, four women and two children, the community started by ceremonially burning the deeds to the land to emphasise their hostility to private ownership and individual power. In the early days this land was managed collectively, meals were eaten together and anyone was welcome to come to the community and erect a shack on the land as a home (see Figure 1.5). The communards had a laundry, and although the women were still responsible for domestic chores, they experienced more freedom than was available 'outside'. Women and men worked together on the land, they were able to wear 'rational dress' (a freer, more practical style of clothing than was conventionally accepted) and women could enter free unions, as Nellie Shaw outlined in her history of the community:

> Those of us who were attempting to go the full length of the Tolstoyan teaching welcomed the idea of free unions as a great improvement on legal marriage. In church, according to law, a woman becomes a chattel, being ringed and labelled as man's property, losing even her name in marriage – almost her identity. We believed that love between two people was the only reason for such unions ... which were far more likely to be enduring. (Shaw, 1935: 128)

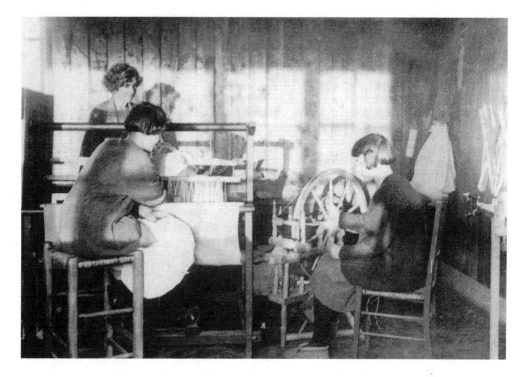

Figure 1.5 Co-operative spinning and weaving at the Whiteway Colony

Over time, the collective spirit at Whiteway was threatened by visitors who did not labour in return for their meals, by those who tested the goodwill of the settlers by taking their goods, and by hostile reports in the press. *The Daily Express*, for example, declared 'The Colony set out to live the highest and simplest life; it has inevitably degenerated into a return to savagery ... No idea can be given of the indolence and sheer animalism of the Whiteway Arcadia' (quoted in Osman, 1992/93: 69). Commentators attacked the colony for its free unions, for nudity and even for the eccentricity of vegetarianism. By the early years of this century community members opted to till their own patches of land or produce their own goods for market, toning down the early communitarianism of the settlement, and so allowing it to survive. On this more pragmatic basis Whiteway was the only early Tolstoyan community to survive over the long term, and the wooden houses and haphazard development of the settlement are still in evidence today.

DiY protest and party: Anarchism reworked in the present

Because anarchism is in its essence an anti-dogmatic and unstructured cluster of related attitudes, which does not depend for its existence on any enduring organization, it can flourish when circumstances are favourable and then, like

> a desert plant, lie dormant for seasons and even years, waiting for the rains
> that will make it burgeon. (Woodcock, 1962: 453)

> [I]f flower power has gone to seed then germination must soon begin. And
> what King Weeds they'll be. (The underground publication *Frendz* [1972],
> quoted in McKay, 1996: 4)

After a long period of relative dormancy, anarchist sentiment resurfaced in the
political rebellions and counter-cultural movements of the 1960s and 1970s.
The surge of libertarianism which swept the world in the wake of civil rights
protests, the anti-Vietnam War movement and demands for women's equality
rekindled the passion for liberty in new times. These new revolutionaries were
alienated by the repressions associated with the Eastern Bloc as much as those
particular to their own governments, and in these circumstances they naturally
tended to develop some form of anarchist philosophy and practice. Established
channels of politics and, in particular, 'left' parties and organisations, were
largely discredited as agents of change, and individuals took matters into their
own hands if they wanted anything done (see Caute 1988; Marshall, 1993).
Rather than replicate the sober politics of the established socialist movement,
the new radicals of the 1960s and 1970s were involved in collective celebration, emphasising the pleasure of rebellion and the rebellious nature of fun:

> The growth of counter-culture, based on individuality, community and joy,
> expressed a profound anarchist sensibility, if not a self-conscious knowledge.
> Once again, it became realistic to demand the impossible. (Marshall, 1993: xi)

Politics was about self-expression ('finding yourself'), autonomy and creativity.
And remarkably, such characteristics are still central to much contemporary
political protest and organisation.

The 1960s laid the foundations of what is sometimes called *DiY culture*.
For example, in Britain, a collection of diverse groups organise direct action
over issues such as the arms industry, road building, land ownership, housing,
hunting, animal rights and even the right to party in peace. There has been an
explosion of autonomous, creative, political organisations in nineties Britain,
all of which prioritise self-activity and expression:

> In the eighties, a lot of people who were hacked off with the way we were
> living, or were just plain bored, got off their arses and did something about
> it ... DIY culture was born when people got together and realized that the only
> way forward was to do things for themselves ... Ingenuity and imagination are
> the key ingredients ... Free parties, squat culture, the traveller movement
> and later Acid House parties pay testament to the energy and vision of people
> who decided it was now time to take their destinies into their own hands.
> (Cosmo, DiY activist, quoted in McKay, 1998: 2)

Ironically, such initiatives have been united by the actions of the last Conservative government, and in particular by the Criminal Justice Act which was
passed in 1994. Designed to stamp out the disparate protests that were delay-
ing road-building programmes or interfering with rural hunts, and to stop the

travellers, ramblers and ravers who were upsetting rural landowners, this legislation merely fanned the flames of protest and collective indignation. The
nationwide response it provoked brought DiY culture into focus, illuminating
the common ground between different groups of activists in different parts of
the country (McKay, 1996). And as a result, there is now a self-conscious network of activists across the UK, as the directory of this new infrastructure of
protest declares:

> The Book is symptomatic of a new awareness, a force of empathy, wit, vision
> and community spirit which has given a fresh sense of empowerment and
> freedom. The scapegoated have become united like never before. The old
> channels of protest and party politics are dead. DiY culture is creating homes
> and entertainment by the people for the people captured in the philosophy
> Deeds Not Words. The Criminal Justice Act has, unintentionally, opened
> doors which can never be shut again. It has motivated the largest direct action
> movement in years. We bring you this message:
>   You are only accountable to yourself in this life, and all you have to believe
> is that you can make a difference. Believe in change. Get a vibe going with others
> who believe the same. REALISE IT realise that making it happen is worth more
> than any amount of revolutionary hypothesizing. (*The Book*, 1995: preface)

Rather than establishing firm, hierarchical organisations, DiYers prefer the network as an organisational form, rejecting the traditional format of meetings,
committees and titles. As Colin Ward explained as early as 1972, anarchist
groups are to be:

> ... *voluntary, functional, temporary* and *small*. They depend, not on
> membership cards, votes, a special leadership and a herd of inactive followers
> but on small, functional groups which ebb and flow, group and regroup,
> according to the task in hand. They are networks not pyramids. (Ward, 1972:
> 137–138; emphasis in original)

For our purposes, it is clear that DiY culture has many continuities with the
anarchist traditions we associate with the last century. Contemporary protest
focuses on *the action of the individual* rather than looking to hierarchical
organisations that are tied up in established channels of protest; it focuses on
the immediate, on action rather than words; and it is associated with *new
social spaces* of organisation – even if they are only in existence for the time
taken to rave or for the months involved in a road-building programme. Just
as the pioneers of Whiteway or the Spanish anarchists sought to do, individuals
are taking direct action to build an alternative social order in the here-and-
now, seeking to undermine the established order by example and lifestyle.
Indeed, many New Age travellers are very explicit about the politics of their
lifestyle, using pagan, rather than Christian, language to echo the Tolstoyan
communitarians who were active 100 years ago in Britain:

> We don't want to resist – that's negative, we want to create ... Similar to direct
> action where 20 people in a rural hamlet can achieve more than 50,000 on
> a march in London can, we believe that lots of people doing lots of small

> gatherings is just as effective in creating social and political change ...
> Empowering individuals and re-connecting them with the land, whilst largely
> going un-noticed by the state. (Laugh, on the Freedom Trail in the West
> Country, quoted in McKay, 1998: 9)

Just as capturing space was central to anarchist experiment in the previous
century, so too it is a feature of contemporary DiY culture. Travellers, eco-
protesters, ravers and squatters all need their own space to live their alterna-
tives, and all have made imaginative use of free space. Hakim Bey (1991) refers
to such anarchistic spaces, where freedom and thus transgression is possible,
as *Temporary Autonomous Zones* (TAZ), and they last as long as the possibilities of the action determine. While urban squatters may be able to forge a
long-term community providing new spaces for cafés and parties, the road
protests and free festivals have tended to have shorter life-spans. This spatial
politics has its roots in the 1960s and 1970s when squatting took off in urban
centres and the free festivals created the space for experiment in rural locations.
Open-air free festivals developed in the wake of the Woodstock festival in the
USA in August 1969, and in Britain, Ubi Dwyer was inspired to co-ordinate
the annual Windsor Free Festivals between 1972 and 1974, as George McKay
relates:

> In the early seventies Ubi was living in a commune in a squatted fire station
> in Fleet Street in London. Tripping on acid in Windsor Great Park he had a
> Blakean vision of a communitarian utopia, which he thought he could bring
> to life by holding 'a giant festival in the grandest park in the kingdom, seven
> miles long!'. Why hold it at Windsor? Because it's an effort to reclaim land
> enclosed for hunting by royalty centuries before – an updating of seventeenth
> century Digger strategy, challenging the later seizure by George III of Windsor
> common land. Ubi strikes at the heart of the British establishment and
> property-owning classes. (McKay, 1996: 16)

Similar events were held in East Anglia, at Glastonbury (the longest survivor
of this phenomenon), and most famously at Stonehenge (annually between
1974 and 1985). And although many of those attending were not explicitly
political, there are clear links with an earlier anarchist tradition, particularly in
the celebration of freedom and the revolutionary overthrow of everyday life:

> Free festivals are practical demonstrations of what society could be like all
> the time: miniature utopias of joy and communal awareness rising for a few
> days from the grey morass of mundane, inhibited, paranoid and repressive
> everyday existence ... The most lively [people] escape geographically and
> physically to the 'Never Never Land' of a free festival where they become
> citizens, indeed rulers, in a new reality. (From an anonymous leaflet, quoted
> in McKay, 1996: 15)

As an example of a TAZ in the contemporary period, the Brighton-based
group *Justice?* squatted a derelict courthouse in 1994, as a focus for the campaign against the Criminal Justice Act, turning the space into a place for all
sorts of events from conferences and cafés to parties:

> In just three days the derelict 100-yr old former magistrate's office was transformed into a thriving community centre with cafe, meditation space, crèche and free entertainment for a free people. Over 1000 people came through the doors. (Schnews, 1995, quoted in McKay, 1996: 175)

The road protest movement has also used space in creative new ways. In opposition to both the M11 and M77, DiYers declared free states – of Wanstonia, Leytonstonia and Pollok – issuing new passports and flags as symbols of the free space (McKay, 1996; Routledge, 1997). Indeed, the earth itself has become a focus of political protest as *The Land is Ours* campaign raises questions about the ownership, development, stewardship and use of the land (see Box 1.9).

Box 1.9 The Land is Ours campaign

The Land is Ours (TLIO) campaign began with an occupation of disused and set-aside land near St George's Hill, Surrey (which was the site of Winstanley's new settlement during the English Civil War) in 1995. About 600 protesters occupied the land, built a village and gardens, publicising the politics of land ownership, use and distribution. Following another occupation near Watlington, Oxfordshire, later that year, the campaign's greatest success has been in setting up a new community on 13 acres of derelict land in Wandsworth, London, during May 1996. For five and a half months 500 activists made their new homes on land that was destined to be home to the ninth large superstore within a radius of a mile and a half. This new urban village of self-erected homes, small gardens and communal spaces attracted an enormous amount of publicity and allowed the campaigners to highlight the destruction of urban communities, the lack of affordable housing and the environmental failures of planning regulations.

TLIO aims to revive popular anger about land ownership, stewardship and development in the UK, as demonstrated in the following passage written by George Monbiot:

> If ordinary people don't like a local authority's decision to approve a development, there's nothing whatsoever that they can do about it. If developers don't like the council's decision to reject their proposed developments, they can appeal to the Secretary of State for the Environment. Developers know that an appeal will cost the council hundreds of thousands to contest. Time and again developers use the threat of appeal as a stick to wave over the council's head, and as often as not the blackmail works. If the council has enough money to fight an appeal, however, and if at appeal the Secretary of State rejects that developers' plans, all they need to do is to submit an almost identical planning application, and the whole process starts again. This can go on until both the money and the willpower of the council and local people are exhausted and the developers get what they want. If the blackmail and extortion still don't work, however, the developers have yet another weapon in their armoury. Planners call it 'offsite planning gain'. You and I would recognise it as bribery. Developers can offer as much money as they like to a local authority, to persuade it to accept their plans. You don't like my high-rise multiplex hypermarket ziggurat? Here's a million quid. What do you think of it now?

Box 1.9 *continued*

> The results of this democratic deficit are visible all over our cities. Where we need affordable, inclusive housing, we get luxury, exclusive estates; where we need open spaces, we get more and more empty office blocks; where we need local trade, we get superstores (and I can confidently predict that in ten years' time there'll be as much surplus superstore space as there is surplus office space today). These developments characteristically generate huge amounts of traffic. Affordable housing is pushed out into the countryside. Communities lose the resources which hold them together.
>
> But if this suspension of accountability is onerous in the towns, it is perhaps even more poignant in the countryside. There the message, with a few exceptions, is clear: It's my land, and I can do what I want with it. (Monbiot, 1998: 182–183)

The land itself has been privatised, and in combination with intensive farming techniques, Monbiot chronicles the loss of hedgerows, habitats, land forms and communities based on the land: 'Features that persisted for thousands of years, that *place* us in our land, are destroyed in a matter of moments for the sake of crops that nobody wants. Our sense of belonging, our sense of continuity, our sense of place, are erased' (*op. cit.*, 184, emphasis in the original). The TLIO campaign, in common with other aspects of DiY culture and protest, is designed to use direct action to return the land to the people, and in so doing, change the way that we live.

The liberation of particular spaces for social and political experiment has also provided a place for art to play a part in popular protest. Echoing the politics of Dadaism and the urban adventures of the Situationist International earlier on in this century (see Box 1.10), DiYers make a point of fusing politics with art as part of their lifestyle:

> So it was time to develop new creative political methods, using direct action, performance art, sculpture and installation and armed with faxes, modems, computers and video cameras. A new breed of 'artist activist' emerged whose motto could well have been creativity, courage and cheek. Their art was not to be about representation but presence; their politics was not about deferring social change to the future but about change now, about immediacy, intuition and imagination. (Jordan, 1998: 132)

Writing about the protest against the extension of the M11 that centred on the battle to preserve Claremont Road, London, Jordan describes how a terrace of 35 homes, with only one of the residents remaining, became a creative space of activism, art and everyday life:

> ... the road – normally a space dominated by the motor car, a space for passing not living, a dead duct between a and b – was reclaimed and turned into a vibrant space in which to live, eat, talk and sleep. Furniture was moved out of the houses into the road, laundry was hung up to dry, chess games were played on a giant painted chess board, snooker tables were installed, fires were lit, a stage was built and parties were held. The 'road' had been turned into a 'street', a street like none other, a street which provided a rare glimpse of utopia, a kind of temporary microcosm of a truly liberated, ecological culture. (Jordan, 1998: 135)

Old cars were painted and upended as art, murals adorned the exterior walls, each house was used creatively as part of the protest, and a huge 100-foot scaffold tower was erected as a local landmark to the campaign. Despite the 'operatic battle' that involved 1300 riot police and bailiffs eventually clearing the 'street' during November 1994, these campaigners sought to carry on their festival on new turf – many of them re-emerging in the activities of the group *Reclaim the Streets*.

Box 1.10 The Situationist International
by David Pinder

The Situationist International (SI) was a revolutionary art and political group that was active between 1957 and 1972, especially in Western Europe. A total of 70 members from 16 different countries participated over those years. Through a range of writings, artworks and radical activities, the situationists confronted what they saw as the increasingly total alienation of everyday life in capitalist and state bureaucratic societies in the post-war period. They believed that, with the extension of commodity relations into all aspects of the social realm, and with the emergence of an image-saturated 'society of the spectacle,' people had become alienated not just from their own labour, but from their creativity, their desires, and their true selves. They therefore advocated creative revolt that could transform everyday spaces and allow people to reclaim control over their own lives. The group's ideas relate to earlier avant-garde interventions, including those associated with Dada and surrealism. They also drew on Marx and Hegel and have connections with currents of western marxism, especially the work of Henri Lefebvre (on marxism and Lefebvre, see Chapter 2). The situationists aimed to reassess the lessons of earlier workers' movements and were sharply critical of most political ideologies, including forms of anarchism. But they shared certain concerns running through this chapter, including the critique of work, the rejection of authoritarianism and the orthodox Left, the interest in autonomy and self-management, and the development of revolutionary contestation. They also looked back to anti-hierarchical struggles such as the Paris Commune of 1871, characterising it as a 'festival' through which everyday life and space were liberated, and as '*the only realisation of a revolutionary urbanism* to date' (in Knabb, 1981: 314–317).

Issues of geography and urban space were central to the SI's project, especially in its early years. Guy Debord, a leading figure in the group, argued in his book *The Society of the Spectacle* of 1967: 'The proletarian revolution is that *critique of human geography* whereby individuals and communities must construct places and events commensurate with the appropriation, no longer just of their labour, but of their total history' (1994, thesis 178). The situationists addressed urbanism, planning and architecture, and the significance of these fields in the reproduction of dominant capitalist interests. They also explored ways of studying and transforming cities through forms of cultural politics and practices of 'psychogeography'. These had been developed earlier in the 1950s by the Lettrist International and included *dérives*, or critical drifts through the streets; and *détournement*, which involved the diversion or hijacking of objects, materials or images and their recomposition to create new effects. The situationists sought to disrupt dominant meanings, to open up

Box 1.10 *continued*

liberating places and routes. Existing cartographies were subverted, and urban spaces remapped according to different experiences, desires and values (see Pinder, 1996). These practices were connected with a utopian vision that called for the transformation of urban space and society in terms of 'unitary urbanism'.

The situationists' notoriety came especially with the revolts in Paris, in May 1968 (see Figures 1.10.1 and 1.10.2). While these are often characterised as a student rebellion, they soon spiralled much wider and led to the occupation of streets and official buildings, including the Sorbonne, and to a general strike in France that involved around 10 million workers. Much of the action was spontaneous and wildcat, taking place outside 'official' political channels such as the trade unions and the French Communist Party, which opposed it. Everyday life and spaces were momentarily transformed, and the state itself was threatened with revolution before its authority was reimposed the following month. Revolts also spread through other parts of the world. The situationists welcomed and sought to fuel these actions as signs of the rejection of contemporary alienation, and the spirit of their spatial critique found expression in graffiti that sprang up around Paris that included: 'Poetry is in the streets!'; 'Under the paving stones, the beach!'. Since then, the situationists have had a wide influence in radical artistic and political circles and increasingly within academia, although their traces have often been unacknowledged. Their critique of human geography has particular resonances with a range of current engagements with the politics of space. It has proved inspiring for a number of those involved in such struggles, being taken up, for example, within elements of DiY protest. (For a selection of situationist writings, see Knabb, 1981; for further introductions, see Bonnett, 1989, 1996; Plant, 1992; Marshall, 1993.)

Figure 1.10.1 Marching in the streets, Paris, 1968

Box 1.10 *continued*

Figure 1.10.2 'The People's University – Yes!', Paris, 1968

Reclaim the Streets seeks to use tarmac as the site for festivals and carnivals within cities, holding 'events' in Camden High Street, London (May 1995, involving 500 people); Upper Street, Islington, London (July 1995, involving 3000 people); the M41, West London (July 1996, involving 10,000 people); Liverpool, to support the sacked dockers (October 1996; also see Chapter 2); and Trafalgar Square for the 'Never Mind the Ballots, Reclaim the Streets' event to mark the General Election and in support of the March for Social Justice (April 1997) (see Figure 1.6). Such events have made it political to dance in the street, subverting authority by having fun and sharing space with strangers: challenging the established order of transportation and road regulation, but also the very mechanics of everyday life. And action has spread to other parts of Britain and beyond, including such places as Hull, Sheffield, Helsinki and Sydney (Jordan, 1998: 150–151).

Anarchism is clearly alive and well: a living and growing part of contemporary political protest in Britain and further afield. DiY culture and protest is reshaping the cartography of organised resistance, forging networks of activists across space and through time; it is finding new ways of using space as a tool in collective defiance; and it is reconfiguring popular ideas about place, the

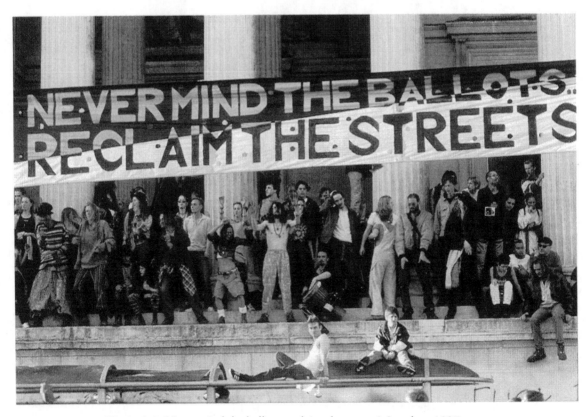

Figure 1.6 'Never mind the ballots, reclaim the streets', London, 1997

land, the rural and the urban. In short, DiY manifests an anarchist geography which, as we will see in the following section, is yet to be reflected in the academy, or within the discipline of geography as it is constructed today.

Anarchism and geography: a disciplinary view

Writing in a prison cell in 1885, Kropotkin argued that the subject of geography is ideally suited to teaching young people to have respect for the natural world and empathy with peoples of other cultures in other parts of the world. In his eyes, geography could be a political weapon, undermining the ideology of imperialist domination prevalent in his day:

> [Geography] must teach us, from our earliest childhood, that we are all brethren, whatever our nationality. In our time of wars, of national self-conceit, of national jealousies and hatreds ably nourished by people who pursue their own egotistic, personal or class interests, geography must be – in so far as the school may do anything to counterbalance hostile influences – a means of dissipating these prejudices and of creating other feelings more worthy of humanity ... It has to enforce on the minds of children that all nationalities are valuable to one another. (Kropotkin, [1885] 1979: 7)

This plea for geography to play a political role in educating people and changing the world, was echoed by a new generation of geographers in the late 1960s. In what later became labelled 'radical geography', groups of geographers organised themselves to remake the discipline as relevant, socially useful and a transformative catalyst in educating minds. These radical geographers argued that the discipline needed to develop new theoretical ideas and new empirical practices to meet the social and environmental challenges of the age. Indeed, the prevailing geographical canon was accused of simply justifying the status quo, explaining existing social and spatial patterns, rather than challenging their very existence. Instead of simply describing the world and its geography, these geographical radicals wanted the discipline to be a tool of revolution, transformation and dissent. And not surprisingly in this context, they looked at the life and work of the anarchists Kropotkin and Réclus for new inspiration.

In his reflection on the legacy of anarchist geography, Richard Peet (1975: 43) declared that, for much of the twentieth century, 'We [Geography] became a discipline for scientifically justifying patterns of social and spatial development based on human competition, human selfishness, and human inequality'. And for Peet, anarchism could form the basis of a new approach to geographical theory, research and teaching. Radical geography could 'take Kropotkin's work as its new beginning, to take his view of human nature as its inherent assumption, to listen to his plea for the widest extension of mutual support, to use our very practical skills to show how mutual aid can be reasserted as the organizing force of future history' (Peet, 1975: 43). Anarchist geography was to be embraced as part of the new radicalism of the late 1960s and 1970s and it was to be the foundation of a new approach to theory, research, teaching and argumentation.

In the event, however, *Antipode*, the journal of radical geography established by this group of activists, illustrates the way in which anarchism was rather sidelined as radical theory developed. By the early 1970s the journal articles indicate that marxism, and later feminism, became the favoured sources of new ideas in research. And indeed, the following chapters in this book outline the real theoretical advances that have been made in both marxist and feminist geography, while in contrast, anarchist geography has not been developed in any real depth. Despite the important geographical implications of anarchist ideas and practice – in taking mutual aid as its premise, in valuing nature and the environment, in establishing new forms of community and in using space as a political resource – there has been little theoretical development of anarchism or any detailed empirical examination of anarchic enterprise since the late nineteenth-century work of Kropotkin and Réclus. Many critical scholars probably take some aspects of anarchist thought for granted, but there has been very little development of the tradition within geography in the last 100 years.

In some senses this may be due to the very nature of anarchist thought, which, as we have seen, prioritises anti-authoritarianism, autonomy and decentralisation. Moreover, many contemporary anarchist movements are anti-intellectual as a matter of policy, prioritising action and deed over thought and the labour of theory, meaning that anarchism is perhaps poorly suited to the explanatory theorising upon which academic human geography has come to depend. Anarchism does not claim to have all the answers; it celebrates the local, the immediate and the deed, making it difficult to deploy it as an explanatory tool in academic endeavour. Yet as many people have questioned the basis of so-called 'grand theory' and any claims to universal explanatory ideas, we might expect anarchism to come into its own (for some preliminary ideas in this regard, see Routledge, 1997). Indeed, as the left, and marxism in particular, have fallen from favour in the years since the 1970s, anarchism might be the source of new dissenting ideas. Such has been the case in the world beyond the academy – as we have seen in the case of DiY culture – and it might be that academic theory and research slowly comes to reflect these new forms of protest and the ideas upon which they depend. Anarchism might be expected to take greater prominence in fermenting dissent in the future, and so attract more interest in the social sciences and geography as a result (the work of Murray Bookchin, 1992, 1995 may come to prominence in this endeavour). There is certainly scope to apply and develop the anarchist tradition within geography, building upon the roots laid in the nineteenth century to make the discipline a force for social change in the twenty-first century.

Conclusions

Anarchists are libertarians who oppose the centralisation of power in institutions such as the Church, state, factory and family. Arguing that a lack of individual autonomy stifles human creativity and sociability, anarchists seek to build a new society in which people live in small-scale, self-governing communities where decision making is shared and the local environment is protected. As part of their action for change, anarchists have involved themselves in co-operatives, new communities and communes, living alternative lifestyles as a strategy to undermine the state and capitalist society from within. In this regard, anarchist dissent focuses on the action of the individual in the here-and-now, and, in so doing, anarchists have collectively used space in creative ways to achieve social change. In this chapter we have examined geographies of anarchist experiment, focusing on the uneven distribution of anarchist ideas and practice; on anarchist settlements in Britain; and on the growth of anarchist sentiment in political protest since the late 1960s and 1970s. A brief overview of the development of DiY culture in Britain during the 1990s illustrated the extent to which anarchism still shapes the landscape of protest as a new generation take direct action to protect the environment, resist the arms industry, defend animal rights and hold parties. Such contemporary campaigns use space as an arena for experiment in new forms of social organisation, keeping the flame of anarchist ideas alive in the present.

Within the discipline of geography, anarchist ideas influenced a new generation of radical scholars who questioned the received wisdoms and hierarchies of the academy during the late 1960s and 1970s. However, as yet, there has been little new theoretical development or application of anarchist ideas in geography. The anarchist tradition might be expected to take greater prominence in fermenting dissent and in stimulating dissident geographies in the future.

Notes

1. The film *Land and Freedom* directed by Ken Loach gives a real flavour of this remarkable period in anarchist history.
2. These settlements were the latest wave of alternative communities established in the UK during the nineteenth century, coming after those inspired by utopian socialism, agrarian socialism and religious sectarianism described by Hardy (1979: 16) and illustrated in Figure 1.4.

References

AMSDEN, J. (1978) Industrial collectivization under workers control: Catalonia, 1936–1939. *Antipode*, **10**: 99–114.

ARMYTAGE, W.H. (1961) *Heavens Below: Utopian experiments in England 1560–1690*. London: Routledge and Kegan Paul.

BEY, H. (1991) *TAZ: the temporary autonomous zone, ontological anarchy, poetic terrorism*. Brooklyn, NY: Autonomedia.

BONNETT, A. (1989) Situationism, geography and poststructuralism. *Environment and Planning D: Society and Space*, 7: 131–146.

BONNETT, A. (1996) The situationist legacy. In S. Home (ed.), *What is Situationism? A Reader*. Edinburgh: AK Press, 192–201.

BOOKCHIN, M. (1992) *From urbanization to cities: towards a new politics of citizenship*. London: Cassell.

BOOKCHIN, M. (1995) *The Philosophy of Social Ecology: essays in diolectical naturalism*. Montreal: Black Rose Books.

BREITBART, M. (1975) Impressions of an anarchist landscape. *Antipode*, 7: 44–49.

BREITBART, M. (1978) Spanish anarchism: an introductory essay. *Antipode*, 10: 60–70.

BREITBART, M. (1981) Peter Kropotkin, the anarchist geographer. In D. Stoddard (ed.), *Geography, Ideology and Social Concern*. Oxford: Blackwell, 134–153.

CAUTE, D. (1988) *Sixty-eight: the year of the barricades*. London: Paladin.

COOK, I. (1990) Kropotkin, prince of geographers. In I. Cook and D. Pepper (eds), *Anarchism and Geography. Contemporary Issues in Geography and Education*, 3(2): 22–26.

COOK, I. and PEPPER, D. (eds) (1990) *Anarchism and Geography. Contemporary Issues in Geography and Education*, 3(2).

DEBORD, G. (1994) [1967] *The Society of the Spectacle*, translated by Donald Nicholson-Smith. New York: Zone Books.

DOLGOFF, S. (ed.) (1980) *Bakunin: on anarchism*. Montréal: Black Rose Books.

DUNBAR, G. (1978) Elisée Réclus, geographer and anarchist. *Antipode*, 10: 16–21.

DUNBAR, G. (1981) Elisée Réclus, an anarchist in geography. In D. Stoddard (ed.), *Geography, Ideology and Social Concern*. Oxford: Blackwell, 154–164.

FLEMING, M. (1988) *Geography of Freedom: the odyssey of Elisée Réclus*. Montréal: Black Rose Books.

FRASER, R. (1994) [1979] *Blood of Spain: an oral history of the Spanish Civil War*. London: Pimlico.

GALOIS, B. (1976) Ideology and the idea of nature: the case of Peter Kropotkin. *Antipode*, 8: 1–16.

GARCIA-RAMON, M.D. (1978) Shaping of a rural anarchist landscape: contributions from Spanish anarchist theory. *Antipode*, 10: 71–82.

GOLDMAN, E. (1969) [1910] *Anarchism and Other Essays*. New York: Dover.

GUILLAUME, J. (1980) Michael Bakunin: a biographical sketch. In S. Dolgoff (ed.), *Bakunin: on anarchism*. Montréal: Black Rose Books, 22–52.

HARDY, D. (1979) *Alternative Communities in Nineteenth Century England*. London: Longman.

HARDY, D. (1990) The anarchist alternative: a history of community experiments in Britain. In I. Cook and D. Pepper (eds), *Anarchism and Geography. Contemporary Issues in Geography and Education*, 3(2): 35–51.

HARPER, C. (1987) *Anarchy: a graphic guide*. London: Camden Press.

JORDAN, J. (1998) The art of necessity: the subversive imagination of anti-road protest and Reclaim the Streets. In G. McKay (ed.), *DiY Culture: party and protest in nineties Britain*. London: Verso, 129–151.

KNABB, K. (ed.) (1981) *Situationist International Anthology*. Berkeley, CA: Bureau of Public Secrets.

KROPOTKIN, P. (1962) [1899] *Memoirs of a Revolutionist*. London: The Cresset Library.

KROPOTKIN, P. (1970) *Kropotkin's Revolutionary Pamphlets*, edited by R. N. Baldwin. New York: Dover.

KROPOTKIN, P. (1979) [1885] What geography ought to be. *Antipode*, **10**: 6–15.

KROPOTKIN, P. (1985) [1899] *Fields, Factories and Workshops Tomorrow*. London: Freedom Press.

KROPOTKIN, P. (1987) [1902] *Mutual Aid: a factor in evolution*. London: Freedom Press.

MALATESTA, E. (1974) [1891] *Anarchy*. London: Freedom Press.

MARSHALL, P. (1993) *Demanding the Impossible: a history of anarchism*. London: Fontana.

MARX, K. (1950) [1847] The poverty of philosophy. In K. Marx and F. Engels, *Selected Correspondence*. Moscow: Foreign Language Publishing House.

MCKAY, G. (1996) *Senseless Acts of Beauty: cultures of resistance since the sixties*. London: Verso.

MCKAY, G. (ed.) (1998) *DiY Culture: party and protest in nineties Britain*. London: Verso.

MILLER, D. (1976) *Kropotkin*. Chicago: University of Chicago Press.

MILLER, D. (1984) *Anarchism*. London: Dent.

MONBIOT, G. (1998) Reclaim the fields and the country lanes! The land is ours campaign. In G. McKay (ed.), *DiY Culture: party and protest in nineties Britain*. London: Verso, 174–186.

ORWELL, G. (1989) [1938] *Homage to Catalonia and Looking Back on the Spanish War*. London: Penguin.

OSMAN, C. (1992/93) Whiteway: the anarchist arcadia. *Diggers and Dreamers*, **92/93**: 62–70.

PEET, R. (1975) For Kropotkin. *Antipode*, 7: 42–43.

PINDER, D. (1996) Subverting cartography: the situationists and maps of the city. *Environment and Planning A*, **28**: 405–427.

PLANT, S. (1992) *The Most Radical Gesture: the Situationist International in a postmodern age*. London: Routledge.

RIGBY, A. (1990) Lessons from anarchist communities. In I. Cook and D. Pepper (eds), *Anarchism and Geography. Contemporary Issues in Geography and Education*, 3(2): 52–62.

ROUTLEDGE, P. (1997) The imagineering of resistance: Pollok Free State and the practice of postmodernism. *Transactions of the Institute of British Geographers* (NS), **22**: 359–376.

SHAW, N. (1935) *Whiteway: a colony in the Cotswolds*. London: C.W. Daniel.

STODDART, D. (1975) Kropotkin, Réclus and 'relevant' geography. *Area*, 7: 188–190.

THOMPSON, E.P. (1994) *Witness Against the Beast: William Blake and the moral law*. Cambridge: Cambridge University Press.

WARD, C. (1972) *Anarchy in Action*. London: George Allen and Unwin.

WOODCOCK, G. (1956) *P. J. Proudhon*. London: Routledge and Kegan Paul.

WOODCOCK, G. (1962) *Anarchism: a history of libertarian ideas and movements*. London: Penguin.

WOODCOCK, G. (ed.) (1977) *The Anarchist Reader*. Hassocks, Sussex: Harvester Press.

2

Class, Capital and Space: Marxist Geographies

Marxism

Marxism is best thought of as a rich tradition of thought which has many branches, each associated with its own time, place and perspective. As the French marxist, Henri Lefebvre (1988: 75) put it: 'There is not *one* Marxism but rather many Marxist tendencies, schools, trends and research projects. Marxism in France does not have the same orientation it has in Germany, Italy, the Soviet Union or China.' In this chapter we look at the roots of this tradition in the work of Karl Marx and Frederick Engels before outlining some of the branches of theory and practice that have evolved since their death. As will be demonstrated, the marxist tradition has enormous importance for understanding the geography of capitalism and the political geography of protest, but it is significant that marxism only arrived in the geographical discipline during the 1970s. This was a time of political upheaval and social protest, and a new generation of students and scholars drew upon marxism as a source of radical ideas to understand and change the world as they found it. As such dreams have rescinded during the 1980s and 1990s, interest in marxism has likewise declined, and there are many branches of marxist analysis and practice which have yet to be translated into geographical thought. This present demise notwithstanding, marxism has had a profound impact upon the development of human geography, and in this chapter, we highlight the contemporary resonances of the tradition.

Marx and the classical tradition

> No thinker in the nineteenth century has had so direct, deliberate and powerful an influence upon mankind as Karl Marx. Both during his lifetime and after it he exercised an intellectual and moral ascendancy over his followers. (Berlin, [1939] 1963: 1)

The ideas of Karl Marx have had an enormous impact on real historical and geographical developments during the nineteenth and twentieth centuries. Marx's writings are still read across the world, they have informed revolutionary practice and they have shaped millions of individual lives. Yet during his lifetime Marx spent much of his time in relative obscurity, he was unable to publish many of his writings, and for long periods of time he took no active part in political organisation (see Box 2.1 for a chronology of his life and work, and see Figure 2.1). Born in Trier, Germany, in May 1818, Marx actually spent most of his life in political exile in London and his family suffered long periods of desperate hardship. When Marx died in March 1883, at the age of 65, he had already lost four of his seven children, three of them during the early years of their lives. Engels gave the Marx family as much financial support as he could, and apart from occasional journalistic work, Marx devoted his life to the development of his ideas. The extensive body of work that Marx produced bears testament to the extraordinary range of his reading, the depth of his scholarship and the clarity of his thinking. Marx's early philosophical work was gradually augmented by historical, political and economic researches which culminated in the three published volumes of *Capital*. More than 100 years after they were written from research notes taken in the British Library in London, these texts still comprise a comprehensive and devastating critique of the capitalist economic order and the social relations on which it depends.

Box 2.1 A chronology of the life and work of Karl Marx

	Writings completed*	Circumstances
1818		Born in Trier, Germany. Father a legal official.
1835		Bonn University to study law.
1836		Berlin University to study jurisprudence, history and philosophy. Joined Young Hegelians.
1841		Awarded Doctorate for thesis entitled *Difference between the Democritean and Epicurean Philosophy of Nature*.
1842		Moved to Cologne as contributor and then editor of the *Rheinische Zeitung*.
1843	**Towards the Critique of the Hegelian Philosophy of Right; On the Jewish Question**	Marries Jenny von Westphalen and moved to Paris when *Rheinische Zeitung* suppressed for radicalism.
1844	Economic and Philosophical Manuscripts; **The Holy Family**	Birth of daughter, Jenny. Meets Engels.
1845	Theses on Feuerbach	Moved to Brussels when expelled from France. Birth of daughter, Laura.

Box 2.1 *continued*

1846	The German Ideology	Set up the Communist Correspondence Committee with Engels. Birth of son, Edgar.
1847	**The Poverty of Philosophy**	Joined the Communist League (used to be League of the Just).
1848	**The Communist Manifesto**	Revolutionary fervour sweeps Europe and Marx expelled from Brussels. Returns to Germany. Editor of *Neue Rheinische Zeitung.*
1849	**Wage Labour and Capital**	Paper suppressed and Marx expelled from Germany. Moved to London. Birth of son, Guido.
1850	**The Class Struggles in France**	Revived Communist League and issued new journal, the *Neue Rheinische Zeitung: Politisch-Oekonomisch Revue.* Death of son, Guido.
1851		Birth of daughter, Franziska and son, Frederick (by Hélène Demuth).
1852	**The Eighteenth Brumaire of Louis Bonaparte**	Disbanded the Communist League. Correspondent for *The New York Tribune*. Death of daughter, Franziska.
1855		Birth of daughter, Eleanor. Death of son, Edgar.
1857	Outline of a Critique of Political Economy (from draft later published as the Grundrisse)	
1859	**Preface to a Critique of Political Economy**	
1862	Theories of Surplus Value	
1863	Capital (vol. II)	
1864	**Inaugural Address to the First International**; Capital (vol. III)	Took part in the launch of the International Working Men's Association (IWMA), St Martin's Hall, London.
1865	Wages, Price and Profit	Conference of the International in London.
1867	**Capital (vol. I)**	
1871	**The Civil War in France**	Paris Commune and conference of the International in London.

Box 2.1 *continued*

1872		IWMA disbanded due to political divisions (officially dissolved in 1876).
1875	**Critique of the Gotha Programme**	
1877	**Anti-Duhring** (a contribution to Engels's work)	
1881		Death of wife, Jenny.
1882		Death of daughter, Jenny.
1883		Death on 14 March, aged 65, London.

* Those texts which were published in Marx's lifetime are shown in bold.
Sources: McLellan, 1973; Mehring, [1918] 1936; Korsch (1938); Callinicos, 1983

Yet *Capital* is not the best place to start getting to grips with Karl Marx. The language and complexity of the argumentation of *Capital* can turn you off very quickly, and in this introduction we use *The Communist Manifesto* as an easier entrée to the ideas of Karl Marx. Published in 1848 as the Manifesto of the League of the Communists, a secret society of largely German artisans, this document is written in a polemical and accessible way (Box 2.2 provides a flavour of the way scholars have read this document over the years). Reading it

Figure 2.1 Karl Marx

KARL MARX, CHEF DE L'INTERNATIONALE.
D'après une photographie de M. Wunder, à Hanovre.

Box 2.2 On reading *The Communist Manifesto*

Irrefutable in its fundamental truths and instructive even in its errors, *The Communist Manifesto* has become a historic document of world-wide significance and the battle-cry with which it closes still re-echoes throughout history: 'Workers of the World, Unite!' (Mehring, [1918] 1936: 151)

The greatest of all socialist pamphlets. No other modern political movement or cause can claim to have produced anything comparable with it in eloquence or power. It is a document of prodigious dramatic force; in form it is an edifice of bold and arresting historical generalizations, mounting to a denunciation of the existing order in the name of the avenging forces of the future. (Berlin, [1939] 1963: 164)

The central drama for which the Manifesto is famous is the development of the modern bourgeoisie and proletariat, and the struggle between them. But we can find a play going on within this play, a struggle inside the author's consciousness over what is really going on and what the larger struggle means. We might describe this conflict as a tension between Marx's 'solid' and his 'melting' visions of modern life. (Berman, 1983: 90)

Marx is not only describing but evoking and enacting the desperate pace and frantic rhythm that capitalism imparts to every facet of modern life. He makes us feel that we are part of the action, drawn into the stream, hurtled along, out of control, at once dazzled and menaced by the onward rush. After a few pages of this, we are exhilarated but perplexed; we find that the solid social formations around us have melted away. By the time Marx's proletarians finally appear, the world stage on which they were supposed to play their part has disintegrated and metamorphosed into something unrecognizable, surreal, a mobile construction that shifts and changes shape under the players' feet. (Berman, 1983: 91–92)

In short, what might in 1848 have struck an uncommitted reader as revolutionary rhetoric – or, at best, as plausible prediction – can now be read as a concise characterization of capitalism at the end of the twentieth century. Of what other document of the 1840s can this be said? (Hobsbawm, 1998: 18)

150 years after its first publication, the *Manifesto* necessarily reflects the world of its time. Indeed, Marx and Engels acknowledged as much in their preface to the German edition of 1872, when they evaluated the lessons of the intervening 25 years:

> In view of the gigantic strides of modern industry in the last twenty-five years, and of the accompanying improved and extended party organisation of the working class, in view of the practical experience gained, first in the February revolution [of 1848], and then, still more, in the Paris Commune [in 1871], where the proletariat for the first time held political power for two whole months, this programme has in some details become antiquated. (Marx and Engels, [1872] 1971: 12)

Yet despite the passage of time, the *Manifesto* still has tremendous power and vision, and for our purposes, it is an excellent introduction to the key ideas of

Karl Marx. In the following text we use the *Manifesto* to highlight (i) the importance of class and class struggle, (ii) the internationalism of capital and labour, (iii) materialism as an explanation for history and ideas, and (iv) the need for a new type of society.

Class and class struggle

Section I of the *Manifesto* opens with the famous words:

> The history of all hitherto existing society is the history of class struggle. Freeman and slave, patrician and plebeian, lord and serf, guild-master and journeyman, in a word, oppressor and oppressed, stood in constant opposition to one another, carried on an uninterrupted, now hidden, now open fight, a fight that each time ended, either in a revolutionary re-constitution of society at large, or in the common ruin of the contending classes. (Marx and Engels, [1848] 1971: 35)

Divisions of class are assigned central importance in marxist understandings of society in general, and of capitalist society in particular. But as this quotation indicates, Marx and Engels also believed class divisions to be a key motor of change in society as the struggles between classes generate new developments, even if the outcomes of such clashes cannot be predicted before they commence. In capitalist society the key class divisions are argued to be between the working and ruling classes, whereby those without economic independence or the means to work for themselves are forced to work for those who own the factories, farms and offices in their location. As the *Manifesto* continues:

> In proportion as the bourgeoisie, i.e. capital, is developed, in the same proportion is the proletariat, the modern working class, developed – a class of labourers, who live only so long as they find work, and who find work only so long as their labour increases capital. These labourers, who must sell themselves piecemeal, are a commodity, like every other article of commerce, and consequently exposed to all the vicissitudes of competition, to all the fluctuations of the market. (Marx and Engels, [1848] 1971: 42)

Class is thus an intimate relationship at the heart of capitalist society, involving the mutual development of the bourgeoisie and the working class: the former depending on the latter as a commodity (labour power) in the processes of production, and the latter depending on the former for their very means of existence. Marx placed labour at the centre of the capitalist mode of production, arguing that labour power is the key commodity that allows accumulation to take place (as illustrated in Box 2.3). By combining labour power (an agreed number of hours in the day, week or year, of a worker's life) with the machines and raw materials at the workplace or office, a capitalist can thus generate new products or services which can be sold for a profit.

Box 2.3 The capitalist labour process

$$M_1 - (LP + MP) - P - M_2$$

where M is money, LP is labour power, MP is means of production (machinery, raw materials, premises, etc.) and P is product. If $M_1 < M_2$ the production process has generated profit, and capital has accumulated.

Marx described the capital used to purchase labour power as *variable capital* (because it is invested in the source of value, as labour is the commodity that adds value in the processes of production) and the capital used to purchase the means of production as *constant capital* (because once purchased, these goods do not increase in their value). For Marx, increasing profit thus depended upon making variable capital work as efficiently as possible, maximising the generation of *surplus value* (the amount of value added by labour after a worker has generated the equivalent of their wages).

In *Capital*, Volume I, Marx outlined two methods for such maximisation of surplus value:

(i) by lengthening the working day

> Capital is dead labour which, vampire-like, lives only by sucking living labour, and lives the more, the more labour it sucks. (Marx, [1867] 1961: 233)

(ii) by increasing the productivity of labour through new means of production

> The object of all development of the productiveness of labour, within the limits of capitalist production, is to shorten that part of the working-day during which the workman must labour for his own benefit, and by that very shortening, to lengthen the other part of the day, during which he is at liberty to work gratis for the capitalist. (Marx, [1867] 1961: 321)

Yet capitalism is more complicated than this, and as contemporary workers know only too well, the vagaries of the market and the strategies of competitor firms make economic life profoundly unstable. Exploitation and insecurity prompt workers to organise in their collective interests, a process which is facilitated by the geographical concentration of capital and workers in ever larger urban locations. As Marx and Engels wrote in the *Manifesto*:

> The unceasing improvement of machinery, ever more precarious; the collisions between individual workmen and individual bourgeois take more and more the character of collisions between two classes. Thereupon the workers begin to form combinations (Trades' Unions) against the bourgeoisie; they club together in order to keep up the rate of wages; they found permanent associations in order to keep up the rate of wages; they found permanent associations in order to make provision beforehand for these occasional revolts. Here and there the contest breaks out into revolt. (Marx and Engels, [1848] 1971: 45)

When they wrote these words, modern trade unions were only beginning to develop in industrialising countries such as Britain, and Marx and Engels saw the unions as emblems of workers' self-organisation (see Box 2.4 for further elaboration of the attitudes of Marx and Engels towards trade unions). Moreover,

Marx and Engels argued that such workers' organisations would prove to be the ultimate contradiction of capitalism. As employers assemble workers together in production for profit, so too they unite workers in a common purpose to improve their living conditions, and by ceasing their labour, workers could realise immense power for change. As Section I of the *Manifesto* concludes:

> The advance of industry, whose involuntary promoter is the bourgeoisie, replaces the isolation of the labourers ... The development of Modern Industry, therefore, cuts from under its feet the very foundation on which the bourgeoisie produces and appropriates products. What the bourgeoisie, therefore, produces above all, is its own gravediggers. Its fall and the victory of the proletariat are equally inevitable. (Marx and Engels, [1848] 1971: 48)

In marxist thought, working class people are thus argued to be the key agents of change because of their position at the heart of the capitalist system – workers' position and experience encourages them to organise collectively whilst also affording them the opportunity to bring down the whole edifice of capitalist society. The extent to which workers have realised their role as the gravediggers of capitalism is a subject we return to later on in this chapter.

Box 2.4 Marx, Engels and trade unions

Trade unions are associations of wage earners for the purpose of maintaining or improving the conditions of their working lives. They have always been a feature of the development of wage-labour.

When Marx and Engels wrote *The Communist Manifesto*, unions were still in their infancy, and in many countries, they were illegal. By the end of the nineteenth century, however, powerful trade union organisations had developed across the industrial world, often in association with mass socialist parties. As Lapides (1987: xv) remarks: 'The modern age in class relations had begun'. Reflecting the development of trade union organisation, Marx and Engels altered their approach towards trade unions during their lifetimes. In the early texts (as we have seen in *The Communist Manifesto* of 1848) they took a very positive view of the role played by trade unionism, but by the 1870s and 1880s their writing is replete with scathing hostility, particularly towards British trade union leaders.

Here Richard Hyman summarises the key arguments in the early stage of their thinking:

> First, unions were a natural product of capitalist industry; workers were forced to combine as a defence against wage-cutting and labour-displacing machinery. Second, unions were not ... ineffectual economically; they could prevent employers reducing the price of labour power below its value. But they could not raise wages about this level, and even their defensive power was eroded by the concentration of capital and recurrent economic crises ... Hence third, the limited efficacy of defensive economic action forced workers to organize increasingly on a class-wide basis, to raise political demands, and ultimately to engage in revolutionary class struggle ... Above all else experience in trade unions enlarged workers' self-confidence and class consciousness. (Hyman, 1983: 482)

Box 2.4 *continued*

And yet by the 1870s and 1880s Marx and Engels argued that British workers (and trade unionists in particular) were part of a *labour aristocracy* who had been bought off with spoils from the world supremacy of British capital. Trade unions had become incorporated into the capitalist system. As Engels wrote in 1885:

> The engineers, the carpenters and joiners, the bricklayers, are each of them a power, to the extent that, as in the case of the bricklayers and bricklayers' labourers, they can even successfully resist the introduction of machinery. That their condition has improved remarkably since 1848 there can be no doubt, and the best proof of this is in the fact that for more than fifteen years not only have their employers been with them, but they with their employers, upon exceedingly good terms. They form an aristocracy among the working-class; they have succeeded in enforcing for themselves a relatively comfortable position ... and they are very nice people indeed nowadays to deal with, for any sensible capitalist in particular and for the whole capitalist class in general. The truth is this: during the period of England's industrial monopoly the English working-class have, to a certain extent, shared in the benefits of the monopoly. These benefits were very unequally parcelled out amongst them; the privileged minority pocketed most, but even the great mass had, at least, a temporary share now and then. (Engels, quoted in Lapides, 1987: 131)

In short, trade unions play a complex and contradictory role in capitalist society, and the experience of the nineteenth century suggests that marxist approaches are not fixed in stone. Analysis depended upon the economic conditions, prevailing politics and organisational practices found at the time.

The internationalism of capital and labour

The Communist Manifesto is replete with imagery of the global voracity of capitalism. Marx and Engels understood capitalism to be a dynamic system, and geographical expansion was already a source of additional wealth as new products, new markets and new workers could be brought into the fabric of an emerging global economy. In language that is remarkable in the context of contemporary debates about 'globalisation', the *Manifesto* of 1848 declared that:

> The need of a constantly expanding market for its products chases the bourgeoisie over the whole surface of the globe. It must nestle everywhere, settle everywhere, establish connections everywhere. The bourgeoisie has through its exploitation of the world market given a cosmopolitan character to production and consumption in every country ... In place of the old wants, satisfied by the productions of the country, we find new wants, requiring for their satisfaction the products of distant lands and climes. In place of the old local and national seclusion and self-sufficiency, we have intercourse in every direction, universal inter-dependence of nations. (Marx and Engels, [1848] 1971: 39)

In his draft of the manifesto (entitled *Principles of Communism*) Engels further elaborated the importance of the geographical interconnections which accompany capitalist social relations:

> It has reached the point that a new machine invented today in England, throws millions of workers in China out of work within a year. Large-scale industry has thus brought all the peoples of the earth into relationship with one another, thrown all the small local markets into the world market, prepared the way everywhere for civilisation and progress ... (Engels, in Marx and Engels, [1848] 1971: 82)

Leaving aside the contentious remark that the world market brings civilisation and progress to the underdeveloped parts of the globe, Marx and Engels recognised the importance of the changing geography of capitalism. This is a major topic to which we return later in this chapter, but as far as the *Manifesto* is concerned, Marx and Engels argued that global capitalism laid the seeds for international class struggle, and the end of national divisions and racism between workers.

In Section II of the *Manifesto* Marx and Engels outlined the political strategies and policies they believed communists should pursue within the trade unions and political parties of the working class, affording internationalism a major place in this platform. They argued that 'In the national struggles of the proletarians of the different countries, they [communists] point out and bring to the fore the common interests of the entire proletariat, independently of all nationality' (Marx and Engels, [1848] 1971: 49). As the world market brought the peoples of the globe closer together, so Marx and Engels argued that the new communist society would be forged at an international scale, closing the *Manifesto* with the famous words:

> Let the ruling classes tremble at a Communistic revolution. The proletarians have nothing to lose but their chains. They have a world to win.
> WORKING MEN OF ALL COUNTRIES, UNITE! (Marx and Engels, [1848] 1971: 74)

Materialism as an explanation for history and ideas

As illustrated in the above discussion of internationalism, Marx and Engels understood the economic development of society to lay the foundations for political, social and cultural change. As they stated in the *Manifesto*:

> Does it require deep intuition to comprehend that man's ideas, views and conceptions, in one word, man's consciousness, changes with every change in the conditions of his material existence, in his social relations and in his social life? What else does the history of ideas prove, than that intellectual production changes its character in proportion as material production is changed? (Marx and Engels, [1848] 1971: 57)

Even at its most basic, in simple agricultural production, labour transforms nature, and in so doing, this labour transforms the possibilities of social being. With new products, new surplus and new ways of working, new forms of society can come into being, and with that development, new human relations take shape (see Box 2.5). In the *Manifesto*, Marx and Engels acknowledged the extraordinary dynamism of capitalist production, which, in turn, had profound implications for social relations, culture and, of course, political protest. The drive to accumulate in the context of competition ensures that the capitalist system never stays still, bringing new ways and means of production into being, often in new places and spaces. As Marx and Engels explained:

> The bourgeoisie cannot exist without constantly revolutionising the instruments of production, and thereby the relations of production, and with them the whole relations of society ... Constant revolutionising of production, uninterrupted disturbance of all social conditions, everlasting uncertainty and agitation distinguish the bourgeois epoch from all earlier ones. All fixed, fast-frozen relations, with their train of ancient and venerable prejudices and opinions, are swept away, all new-formed ones become antiquated before they can ossify. All that is solid melts into air, all that is holy is profaned, and man is at last compelled to face ... his real conditions of life, and his relations with his kind. (Marx and Engels, [1848] 1971: 39)

This 'everlasting uncertainty and agitation' is still characteristic of contemporary capitalism and ensures that new forms of production, new markets and new relationships are constantly brought into being. In so doing, capitalism lays the seeds for new forms of political organisation and new ways of thinking about the world (witness the growing impact of the Internet on international relations, political practice and information exchange: see Box 2.9 on page 68). Materialism is thus a key building block of marxist thought. The mode of production and reproduction is understood as the root of ideas, culture and historical development.

Box 2.5 Marx and materialism

Marx argued that the way in which people live and the means of their survival (which for the majority in capitalist society involves wage-labour) will determine the ideas in their heads. As he explained in *The German Ideology* (written with Engels) and *A Contribution to the Critique of Political Economy*:

> Men, developing their material production and their material intercourse, alter, along with this their real existence, their thinking and the products of their thinking. Life is not determined by consciousness, but consciousness by life. (Marx and Engels, [1846] 1970: 47)

> The mode of production of material life conditions the social, political and intellectual life-process in general. It is not the consciousness of men that determines their being, but, on the contrary, their social being that determines their consciousness. (Marx, quoted in McLellan, 1988: 19)

Box 2.5 *continued*

As a result of this analysis, marxists have tended to distinguish between the economic *base* of any society and the political, social and cultural *superstructure* that rises upon the foundations of any particular form of economic organisation. Yet it is important to acknowledge that Marx and Engels did not take a crude deterministic approach to these relations but, rather, argued that material conditions limited the possibilities of developments in consciousness, politics and cultural forms. Had it been otherwise, the political organisation and intervention they advocated would have been a waste of time. As far as consciousness is concerned, then, Marx argued that people could act within the conditions found at that time:

> Men make their own history, but they do not make it just as they please; they do not make it under circumstances chosen by themselves, but under circumstances directly encountered, given and transmitted from the past. (Marx, [1852] 1950: 225)

Moreover, as regards the base and superstructure model developed by Marx, Engels wrote the following in a letter to Joseph Bloch in 1890:

> The economic situation is the basis, but the various elements of the superstucture – political forms of the class struggle and its results ... political, juristic, philosophical theories, religious views ... also exercise their influence upon the course of the historical struggles and in many cases preponderate in determining their *form*. There is an interaction of all these elements, in which, amid all the endless host of accidents ... the economic movement finally asserts itself as necessary. (Engels, quoted in McLellan, 1988: 69)

Despite its many critics, then, historical materialism does not necessarily invoke a crude deterministic reading of social development.

The need for a new type of society

As a political polemic, the *Manifesto* was designed to make the case for a new form of social organisation called communism. The Communist League aimed to achieve 'the overthrow of the bourgeoisie, the rule of the proletariat, the ending of the old society which rests on class contradiction and the establishment of a new society without classes or private property' (quoted in Hobsbawm, 1998: 3), and this was their manifesto. The key plank of this political programme involved the abolition of private property (the root of class division) and the socialisation of wealth. In the new world of communism, workers would no longer labour for the propertied class but would labour to enrich the wider, collective society:

> In bourgeois society, living labour is but a means to increase accumulated labour [capital]. In Communist society, accumulated labour is but a means to widen, to enrich, to promote the existence of the labourer. (Marx and Engels, [1848] 1971: 51)

The *Manifesto* then provides a list of some of the measures which would be enacted when the working class has, through revolution, become the new ruling class (a battle which with the benefit of hindsight is, of course, much more difficult to win than was suggested in 1848), ending with the powerful words:

> In place of the old bourgeois society, with its classes and class
> antagonisms, we shall have an association, in which the free
> development of each is the condition for the free development of all.
> (Marx and Engels, [1848] 1971: 60)

Revolutionaries of the nineteenth and twentieth centuries have invoked these words of Marx and Engels as they fight for a more just and equal society, and we look at some of these movements and experiments later on in this chapter.

The geography of capitalism

In the years after the death of Marx and Engels, a new generation of marxist thinkers and activists came into positions of prominence in the international working class movement. Together with Marx and Engels, this collection of individuals are usually associated with 'classical' marxism (see Box 2.6), in contrast to the ideas of the 'western' marxists who were active in the years before and after the Second World War (a group of thinkers to whom we return later on in this chapter). The second generation of 'classical' marxist thinkers, such as Kautsky, Plekhanov, Lenin, Luxemburg and Trotsky, took on leading positions in revolutionary parties and, as a result, they developed Marx's sketchy ideas about political organisation while also furthering his analysis of the economy (looking at issues such as finance capital, imperialism and the role of the state). This body of marxist thought is thus broadly characterised by a focus on political-economy and on revolutionary practice rooted in the agency of the working class (and, as such, it reflects the 'classical' ideas of Karl Marx). As we will see in this section, classical marxism has a strong geographical current within it, and it is this brand of marxism which has had the greatest influence on the geographical discipline since the 1970s and 1980s.

Box 2.6 The classical marxists

	Dates	Place of birth	Key writings
Karl Marx	1818–1883	Trier, Germany	See Box 2.1 on p 43–5
Frederick Engels	1820–1895	Barmen, Germany	*The Condition of the Working Class in England* (1845); *The Origin of the Family, Private Property and the State* (1884); *Dialectics and Nature* (1927)

Box 2.6 *continued*

Antonio Labriola	1843–1904	Cassino, Italy	*Essays on the Materialist Conception of History* (1895); *Socialism and Philosophy* (1898)
Franz Mehring	1846–1919	Schlawe, Germany	*History of German Social Democracy* (1897–98); *Karl Marx* (a biography) (1918)
Karl Kautsky	1854–1938	Prague, Czech Rep.	*Foundations of Christianity* (1908); *The Road to Power* (1909)
Georgii Plekhanov	1856–1918	Tambov, Russia	*The Development of the Monist View of History* (1894); *The Role of the Individual in History* (1898); *Fundamental Problems of Marxism* (1908)
Vladimir Lenin	1870–1923	Simbirsk, Russia	*Development of Capitalism in Russia* (1899); *What is to be Done?* (1902); *Imperialism: The highest stage of capitalism* (1916); *State and Revolution* (1917)
Rosa Luxemburg	1871–1919	Zamość, Poland	*Social Reform or Revolution* (1898-99); *Mass Strike, Party and Trade Unions* (1906); *The Accumulation of Capital* (1913)
Rudolf Hilferding	1877–1941	Vienna, Austria	*Finance Capital* (1910)
Leon Trotsky	1879–1940	Kherson, Ukraine	*History of the Russian Revolution* (1932–33); *The Revolution Betrayed* (1937); *The Permanent Revolution and Results and Prospects* (1962)
Otto Bauer	1881–1938	Vienna, Austria	*Die Nationalitatenfrage und die Sozialdemokratie* (1907); *Die Österreichische Revolution* (1923)
Evgeny Preobrazhensky	1886–1937	Orel, Russia	*The Crisis of Soviet Industrialization* (1921–27); *From NEP to Socialism* (1922); *The New Economics* (1926)

Box 2.6 *continued*

Nikolai Bukharin	1888–1938	Moscow, Russia	*Imperialism and the World Economy* (1917–18); *Economics of the Transformation Period* (1920); *Historical Materialism: A system of sociology* (1921)

Sources: Anderson, 1979: 8; Bottomore et al., 1983

In *The Communist Manifesto*, Marx and Engels highlighted the way in which the drive to accumulate involves the erosion of spatial barriers to trade, markets and resources at a global scale, and this theme has persisted throughout the classical marxist tradition. It is a matter to which Marx returned in his later political-economic writings, in which he argued that capitalists seek to annihilate space with time, reducing the time it takes to transport raw materials, goods and people across space and so realise a return on investment:

> While capital must on the one side strive to tear down every spatial barrier to intercourse, i.e., to exchange and conquer the whole earth for its market, it strives on the other side to annihilate this space with time ... The more developed the capital ... the more does it strive simultaneously for an even greater extension of the market and for greater annihilation of space by time. (Marx, *Grundrisse*, [1857–58] 1973: 539)

Yet, as the geographer David Harvey (1975) has pointed out, the establishment of transportation links then tends to promote the concentration of capitalist development and urbanisation at key nodes in this network. Spatial extension and agglomeration thus tend to develop alongside one another as the processes of capitalist production and reproduction refigure the landscape:

> Geographical expansion and geographical concentration are both to be regarded as the product of the same striving to create new opportunities for capital accumulation. In general, it appears that the imperative to accumulate produces a concentration of production and of capital at the same time as it creates an expansion of the market for realization. (Harvey, 1975: 12)

Just as Marx understood capitalism to be a dynamic system, changing over time, so too, this dynamism is argued to leave its mark on the geographical landscape as new centres of production and agglomeration come into being and old ones fall into decay. Previous rounds of investment become obstacles to the further development of capitalism and, as new markets are sought for more productive investment, existing nodes of accumulation lose their positions of strength. As Harvey explains:

> The geographical landscape which fixed and immobile capital comprises is both a crowning glory of past capital development and a prison which inhibits

the further progress of accumulation because the very building of
this landscape is antithetical to the 'tearing down of spatial barriers'
and ultimately even to the 'annihilation of space by time'. (Harvey,
1975: 13)

The capitalist mode of production is thus inevitably associated with *uneven
development*, as there will always be opportunities for increased rates of accumulation through the exploitation of new markets, new labour pools, new
technologies and advantageous trade routes. It is this geographical dimension
to the accumulation of capital that Harvey (1982) calls the *spatial fix* – as the
owners of capital can escape low rates of return and even crisis in one part of
the globe by reinvesting their capital elsewhere. Moreover, this geographical
dimension is fundamental to the very evolution of capitalist society, altering the
ways in which production, consumption and exchange take place. In recognition of this spatiality, Harvey reconfigures the concept of historical materialism
to include geography. In his reading, a *historical–geographical materialist* perspective highlights the importance of both space and time in the evolution of
capitalism and its associated social relations.

In her reading of Marx's analysis of capital, Rosa Luxemburg developed
an understanding of the uneven spatial development of capitalism, arguing
that capitalists can only maintain profit rates by exploiting non-capitalist
spaces. In an era of imperialism, Luxemburg witnessed the way in which
European capital was destroying self-sufficient economies in the non-capitalist
world:

> Through destruction of the primitive barter relations in these countries,
> European capital opens the doors to commodity exchange and production,
> transforms the population into customers of capitalist commodities and
> hastens its own accumulation by making mass raids on their natural resources
> and accumulated treasures. (Luxemburg, [1921] 1972: 59)

In *The Accumulation of Capital*, Luxemburg argued that the mass production
and heavy machinery industries in the developed European economies were
able to survive only by penetrating non-capitalist spaces for markets, goods
and labour (and she used examples from countries and continents such as
India, America, Australia, China and Africa). Luxemburg suggested that there
were three phases to this conquest of space: (i) the struggle against the natural
economy, (ii) the introduction of commodity economy, and (iii) the struggle
against the peasant economy. By destroying non-capitalist systems of production and survival through colonial policy, land seizure, trade strategy, taxation
and war, the peoples of the subjugated world were brought into the ambit of
the capitalist world economy and were thus forced to work for a wage in order
to buy the commodities they needed to survive (and here Luxemburg cites the
powerful example of China and the opium wars with the British in 1839–42
and 1856–60; see also Chapter 5 for further discussion of imperialism). In
Luxemburg's vision, then, the world would become progressively more capitalist. As she explained:

> Historically, the accumulation of capital is a kind of metabolism between capitalist economy and those pre-capitalist methods of production without which it cannot go on and which, in this light, it corrodes and assimilates. Thus, capital cannot accumulate without the aid of non-capitalist organisations, nor, on the other hand, can it tolerate their continued existence side and side with itself. Only the continuous and progressive disintegration of non-capitalist organisations makes accumulation of capital possible. (Luxemburg, [1913] 1951: 416)

But as capitalism progressively extended into all the nooks and crannies of the world, Luxemburg predicted that competition between the European powers for control over the remaining non-capitalist spaces would become ever more fierce. Moreover, imperialism would prove to be its own downfall, as its success would ultimately limit the scope for further capitalist profit and growth.

With hindsight, it can be seen that Luxemburg overestimated the ability of capitalism to conquer the whole world while she also underestimated the ability of capital to exploit and augment the spatial differences within and between capitalist nations. It was Lenin who took up the cause of such understanding in his analysis of imperialism. Lenin argued that as capital grew ever more concentrated in cartels, corporations and banks, these institutions (backed by the nation-state of their origin) would compete for territorial control, thus securing markets, resources and labour. The United Kingdom, France, Germany and the United States of America owned 80% of the world's capital by 1910, and it was the first three of these nations, in particular, that led the battle for empire. Using data for the years 1884–90, Lenin noted that:

> England during these years acquired 3.7 million square miles of territory with a population of 57 million; France acquired 3.6 million square miles with a population of 36.5 million; Germany one million square miles with 16.7 million inhabitants; Belgium 900,000 square miles with 30 million inhabitants; Portugal 800,000 square miles with 9 million inhabitants. (Lenin, [1917] 1934: 71)

Yet, unlike Luxemburg, Lenin did not suggest that this conquest of space erodes spatial differentiation across the globe or that it ultimately limits the potential for profitable growth. Indeed, the colonies of empire were often kept in a state of underdevelopment, being used as a source for raw materials without having the means to embark on indigenous industrial growth, while they provided a source of immense wealth to the colonising powers (a theme later elaborated by development theorists who suggested that the powerful keep poor nations in a state of underdevelopment; see Baran, 1957; Frank, 1969; Emmanuel, 1972; see also Blaut, 1975, 1994; Slater, 1977). In this model, uneven development was an inevitable outcome of capitalist accumulation and, although it could provoke war (as signalled by Luxemburg), geographical differentiation would not necessarily lessen as capitalist expansion and development took place.

Within the geographical discipline these ideas have been developed by David Harvey and Neil Smith to suggest that uneven development is the hallmark of the capitalist mode of production. The twin processes of spatial extension and agglomeration are argued to be necessary outcomes of capitalist production, consumption and exchange. As Smith puts it in his introduction to *Uneven Development*:

> Deindustrialization and regional decline, gentrification and extra-metropolitan growth, the industrialization of the Third World and a new international division of labour, intensified nationalism and a new geo-politics of war – these are not separate developments but symptoms of a much deeper transformation in the geography of capitalism. (Smith, [1984] 1990: xi)

> The point is that uneven development is the hallmark of capitalism. It is not just that capitalism fails to develop evenly, that due to accidental and random factors the geographical development of capitalism represents some stochastic deviation from a generally even process. The uneven development of capitalism is structural rather than statistical ... uneven development is the systematic geographical expression of the contradictions inherent in the very constitution and structure of capital. (Smith, [1984] 1990: xiii)

Geographical unevenness is thus inherent in the way in which capitalism develops, as new opportunities for profit are exploited when existing sources of value decline. Of course, these processes are never universal and some capital will tend to remain in place long after other segments have been disinvested (due to market access, labour supply or even commitment to place), but the overall pattern is one of uneven geographical change. Moreover, this creative destruction takes place at a variety of scales, from the urban, regional and national to the global. As Smith explains, capital see-saws across the landscape in pursuit of accumulative advantage:

> If the accumulation of capital entails geographical development and if the direction of this development is guided by the rate of profit, then we can think of the world as a 'profit surface' produced by capital itself ... Capital moves to where the rate of profit is highest (or at least high), and these moves are synchronized with the rhythm of accumulation and crisis. The mobility of capital brings about the development of areas with a high rate of profit and the underdevelopment of those areas where a low rate of profit pertains. (Smith, [1984] 1990: 148–149)

Once investment has taken place in a particular industry or commodity, in a particular region or nation, the rate of return will tend to fall over time as competitors enter the market, as new methods of production are brought on line, and as workers start to organise for better wages and conditions. So then the search for new opportunities (a new spatial fix) begins and the process of uneven development continues.

The geography of working class organisation

While we have addressed the uneven development of capitalism in the previous section, it is important to acknowledge that this geography also has major implications for the way in which resistance takes place. If, as Marx and Engels suggested in *The Communist Manifesto*, working class organisations are to be the harbingers of a new society, then these organisations have to devise a strategy for overcoming the geographical inequalities produced by capitalist investment and infrastructure. Just as it shapes the landscape of capitalism, uneven development leaves its mark on the geography of working class populations, institutions, organisations and political views. Doreen Massey's [1984] (1994) *Spatial Divisions of Labour* provides a pathbreaking overview of the way in which uneven capitalist development leaves its mark on the landscape of employment and class relations (and for further examples of geographical explorations of economy and class, see Dunford and Perrons (1983) and Thrift and Williams (1987)). As Smith states in the conclusion to *Uneven Development*, the struggle against capitalist social relations needs to respond in geographical kind:

> Capitalism has always been a fundamentally geographical project. It may be not too soon to suggest, and I hope not too late, that the revolt against capitalism should itself be 'planning something geographical'. (Smith, [1984] 1990: 178)

While there are many ways in which geographical differentiation can be borne out in political ideas and practice (including divisions based on race, religion, regional origin, urban origin or even housing district), Marx was keenly aware of the ways in which national differences could act as a wedge between different groups of workers (see Box 2.7). As we have seen in *The Communist Manifesto*, Marx and Engels argued that the struggle for socialism had to involve international organisation, and in 1864 Marx took a leading role in the first organisational experiment to foster such transnational links. The inaugural meeting of the International Working Men's Association (the First International) was held in St Martin's Hall, London, on the basis of connections which had been made during the London builders' strikes of 1858–61 (see also Chapter 1 and Figure 2.2). In his address to this meeting Marx asked the delegates to consider the need for internationalism:

> ... all efforts aiming at that great end [emancipation] have hitherto failed from the want of solidarity between the manifold divisions of labour in each country, and from the absence of a fraternal bond of union between the working classes of different countries ... the emancipation of labour is neither a local nor a national, but a social problem, embracing all countries in which modern society exists, and depending for its solution on the concurrence, practical and theoretical, of the most advanced countries. (Marx, [1871] 1950: 350)

Figure 2.2 The inaugural meeting of the First International, London, 1864

> This Association is established to afford a central medium of communication
> and co-operation between working men's societies existing in different
> countries and aiming at the same end, *viz.*, the protection, advancement, and
> complete emancipation of the working classes. (Marx, [1871] 1950: 351)

British trade union leaders who supported the International did so for more
pragmatic reasons, as many of the trades involved were particularly threatened
by outside competition and foreign labour supplies (including tailors, basket
makers and boot makers; Milner, 1990). As one trade union leader explained
to the same meeting:

> We find that whenever we attempt to better our social condition by reducing
> the hours of toil, or by raising the price of labour, our employers threaten
> us with bringing over Frenchmen, Germans, Belgians, and others to do our
> work at a reduced rate of wages, and we are sorry to say this has been done,

Box 2.7 Marx and internationalism

The need to overcome racial and national divisions which divided the working class prompted Marx to take a leading role in the International Working Men's Association (the First International) which was established in London in 1864 (also see Chapter 1).

His attitude towards nationalism and xenophobia is perhaps best demonstrated in his writings on the 'Irish Question' in which Marx argued that nationalism benefited the ruling class to the detriment of working class organisation:

> Every industrial and commercial centre in England now possesses a working class *divided* into two hostile camps, English proletarians and Irish proletarians. The ordinary English worker hates the Irish worker as a competitor who lowers his standard of life. In relation to the Irish worker he feels himself a member of the nation and so turns himself into a tool of the aristocrats and capitalists *against Ireland*, thus strengthening their domination *over himself*. He cherishes religious, social and national prejudices against the Irish worker. His attitude towards him is much the same as that of the 'poor whites' to the 'niggers' in the former slave states of the USA ... This antagonism is artificially kept alive and intensified by the press, the pulpit, the comic papers, in short by all the means at the disposal of the ruling classes. This *antagonism* is the *secret of the impotence of the English working class*, despite its organisation. It is the secret by which the capitalist class maintains its power. And that class is fully aware of it. ('On the Irish Question', *Selected Correspondence*, 236-237, emphasis in the original)

not from any desire on the part of our continental brethren to injure us, but through a want of regular and systematic communication between the industrial classes of all countries. Our aim is to bring up the wages of the ill paid to as near a level as possible with that of those who are better remunerated, and not to allow our employers to play us all one against the other, and to drag us down to the lowest possible condition, suitable to their avaricious bargaining. (Odger's address to French workers, quoted in Lorwin, 1929: 34)

Divorced from Marx's vision of a socialist internationalism, these British trade unionists sought to defend the superior working conditions of their members through international contact, indicating a division between political and pragmatic internationalism.

Between 1864 and 1868 the First International made a number of successful interventions in various labour disputes by appealing for solidarity between nationalities (examples include the assistance given to London wire and Paris bronze workers). Moreover, the First International was successful in establishing networks of trade unionists and political activists within and between various countries. Although the First International had effectively disbanded due to political differences and state repression by 1872, this model of organisation was later used by trade unions and working class parties to set up International Trade Secretariats (ITS) (see Box 2.8) and the Second International (1889–1914).

Box 2.8 The International Trade Secretariats

A number of International Trade Secretariats (ITS) were formed by trade unionists between 1890 and 1910, and by 1914 there were 28 such bodies, bringing together workers from trades as diverse as hat-making, mining and transportation. As Lorwin explains:

> The main effort of these secretariats was to spread information about trade conditions in different countries, to keep members informed about strikes in their trades, to make appeals for financial aid in case of large strikes, to prevent workers of one country from acting as strike breakers in another, and to promote trade unions in the less organised countries. (Lorwin, 1929: 99)

As an example of such development, 33 delegates from the food and drinks trades met in Zürich, Switzerland, in August 1920 to form the International Union of Men and Women Workers in the Food and Drink Trades (IUMWFDT). Representatives from 13 countries, including Germany, Hungary, Italy, Sweden and the US, determined their purpose in the following resolution:

> ... to protect and promote the economic and social interests of all the workers of the association; to strengthen by all available means the international solidarity of the working classes and to support all national and international action in the struggle against the exploitation of labour; to support everywhere the struggle against imperialism and militarism; and to work for the suppression of the capitalist regime by the realisation of the socialist economic system. (International Labour Organisation, 1920: 7)

In addition to this international interconnection on the basis of trade, various national federations of labour also began to forge new alliances. By 1913, 18 national union federations (representing 7 million workers) had affiliated to the International Secretariat of National Trade Union Centres (IS), which involved delegates attending annual meetings, taking part in information exchange and annual reports. As Milner remarks, this new international body did not seek to erode national differences, but to work with them:

> The IS's measures were a product of national organization, in contrast with the days of the First International, when the International's centralized and public attempts to prevent strike-breaking activities had been necessary because of the absence of national organisation. (Milner, 1990: 97)

This national focus did not prevent the co-ordination of solidarity, and in the appeal for Finnish metal workers in 1907–08 considerable sums were raised: 2785 marks from Denmark; 2468 marks from Germany; 1378 marks from Norway; 1043 marks from the Netherlands; 266 marks from Croatia; 206 marks from Austria; and 61 marks from Bulgaria (from Milner, 1990: 102).

The ITS and the IS were shattered by the 1914–18 war, not least because erstwhile comrades found themselves on opposing sides in the trenches. Although both types of organisation have continued after the wars of the twentieth century, the IS evolving into the International Confederation of Free Trade Unions (ICFTU), they were also profoundly affected by the Russian revolution of 1917.

Box 2.8 *continued*

In 1999 the following ITS still organise trade unionists across national borders:
   International Union of Foodworkers (IUF)
   International Metalworkers Federation (IMF)
   Communications International (PTTI)
   International Federation of Commercial, Clerical, Professional and
   Technical Employees (FIET) (see Figure 2.8.1 for a contemporary example
   of international trade union action, led by FIET)
   International Federation of Building and Wood Workers (IFBWW)
   International Federation of Chemical, Energy, Mine and General Workers'
   Unions (ICEM)
   International Federation of Journalists (IFJ)
   International Transport Workers' Federation (ITF)
   Media and Entertainment International (MEI)
   Public Services International (PSI)

INTERNATIONAL JUSTICE FOR

15th JUNE 1998 JANITORS

fiet

Dear Sir Clive,

On behalf of 11 million members of over 400 trade unions in 125 countries organised in **FIET**, I appeal to your company to respect the rights of your 140,000 employees worldwide to justice and dignity.

June 15 is the day when **FIET** affiliates around the world honour the "invisible" workers who maintain and clean offices, industrial properties and residential buildings and who preserve hygiene, safety and security of millions of their clients on a daily basis, often hidden from public eye.

I fully support the action by **FIET** and its affiliates and for all employees in Europe to be represented on a genuine European Works Council.

Yours sincerely,

Name: _____

Union: _____

Country: _____

Sir Clive Thompson
Chairman,
RENTOKIL INITIAL PLC
East Grinstead
West Sussex
RH19 2BR
United Kingdom

Figure 2.8.1 International Justice for Janitors Campaign, led by FIET, 1998

Source: www. ICFTU.org

The marxist tradition has thus sought to bridge the interpersonal barriers which are crafted by the uneven development of capitalism through conscious organisation. However, in the years since Marx's death, the decisive role of uneven political development has become more apparent than ever. This is particularly clear when we look at the experience of the Russian revolution of 1917. Marx would never have predicted that the first socialist revolution would, or could, have occurred in a relatively underdeveloped country such as Russia. From Marx's analysis, socialist revolution would be expected in those places where the working class had been in existence the longest, where the capitalist system was most developed and where workers had strong political organisations (countries such as Britain, Germany and France). In 1917, however, revolution broke out in a place where capitalism had barely begun to take root and where the working class formed a small minority of the population (see Figure 2.3). In his theory of *permanent revolution*, Trotsky sought to

Figure 2.3 Storming the Winter Palace, St Petersburg, 1917

explain this development, and to outline the political implications of uneven-economic development. Trotsky explained the Russian revolution as a manifestation of the process of *combined and uneven development*, whereby countries on the periphery of capitalism were exploited at the same time as they were subjected to the latest production technology through investment made by those in the advanced economies of Europe and the United States:

> Russian industry did not repeat the development of the advanced countries, but inserted itself into this development, adapting their latest achievements to its own backwardness. Just as the economic evolution of Russia as a whole skipped over the epoch of craft-guilds and manufacture, so also the separate branches of industry made a series of special leaps over technical productive stages that had been measured in the West by decades. Thanks to this, Russian industry developed at certain periods with extraordinary speed. (Trotsky, [1930] 1977: 31)

In a poor country such as Russia, the native bourgeoisie were weak and almost half the capital invested was foreign owned. In contrast to the weakness of the ruling class, however, the organisations of the working class did not have the deep reformist, conservative roots which Marx and Engels witnessed in Britain. The working class in Russia grew very quickly, in very large factories, and as Trotsky remarked in *The History of the Russian Revolution*, the Russian workers were 'hospitable to the boldest conclusions of revolutionary thought' (Trotsky, [1930] 1977: 33). As a result, socialist revolution was possible in a country where many marxists would have least expected it to occur, as Trotsky explains:

> The law of combined development here emerges in its extreme expression: starting with the overthrow of a decayed medieval structure, the revolution in the course of a few months placed the proletariat and the Communist Party in power. (Trotsky, [1930] 1977: 35)

The Russian revolution thus signalled the impact of the combined and uneven development of capitalism upon political organisation, but for Trotsky, revolutions in poor, developing countries could survive only if they spread to more advanced and prosperous locations.[1] In *The Permanent Revolution* he suggested that internationalism remained key to building socialism in any location:

> Internationalism is no abstract principle but a theoretical and political reflection of the character of world economy, of the world development of productive forces and the world scale of class struggle. *The socialist revolution begins on national foundations – but it cannot be completed within these foundations.* The maintenance of the proletarian revolution within a national framework can only be a provisional state of affairs, even though, as the experience of the Soviet Union shows, one of long duration. In an isolated proletarian dictatorship, the internal and external contradictions grow inevitably along with the successes achieved. If it remains isolated, the proletarian state must finally fall victim to these contradictions.

> *The way out for it lies only in the victory of the proletariat of the advanced countries. Viewed from this standpoint, a national revolution is not a self-contained whole; it is only a link in the international chain.* The international revolution constitutes a permanent process, despite temporary declines and ebbs. (Trotsky, [1929] 1969: 133; emphasis added)

The isolation of the Russian revolution had a devastating – and, in many cases, life-threatening – effect on the lives of the people, as Stalin secured his control by building 'socialism in one country' and by imposing a repressive system of authoritarian rule. For our purposes, however, it is important to acknowledge that for Lenin and Trotsky, revolution could never survive in one isolated location. Indeed, the Bolshevik Party established the Third International (or Comintern) and the Red International of Labour Unions with a view to overcoming the uneven development of political organisation. In its early days, before Stalinisation, this body aimed to forge international connections between workers' parties and organisations. As Trotsky remarked in *The Permanent Revolution*, the Bolsheviks sought to highlight the connections between workers across national divides:

> If we take Britain and India as polarised varieties of the capitalist type, then we are obliged to say that the internationalism of the British and Indian proletariats does not at all rest on an *identity* of conditions, tasks and methods, but on their indivisible *interdependence*. Successes for the liberation movement in India presuppose a revolutionary movement in Britain and vice versa. Neither in India nor in England is it possible to build an *independent* socialist society. Both of them will have to enter as parts into a higher whole. Upon this and only upon this rests the unshakeable foundation of Marxist internationalism. (Trotsky, [1929] 1969: 150; emphasis in the original)

> Britain's dependence upon India naturally bears a qualitatively different character from India's dependence upon Britain. But this difference is determined, at bottom, by the difference in the respective levels of development of their productive forces, and not at all by the degree of their economic self-sufficiency. India is a colony; Britain, a metropolis. But if Britain were subjected today to an economic blockade, it would perish sooner than would India under a similar blockade. (Trotsky, [1929] 1969: 152)

To counter uneven development, then, internationalism has to bridge the divides between workers. As Smith argues:

> The political future for the working class lies precisely in the equalization of conditions and levels of production, a process continually frustrated within capitalism. This is the real historical resolution of the contradiction between equalization [combination] and differentiation [unevenness]. (Smith, [1984] 1990: 153)

In the contemporary era of globalisation, some labour organisers and political activists are revisiting the history of internationalism with a view to challenging multinational capital at a transnational scale (see Wills, 1998;

MacShane, 1996; Herod, 1995; Moody, 1997). Waterman (1998) has attempted to theorise this 'new labour internationalism', suggesting that the following principles (amongst others) must be central to new ways of organising labour:

- prioritising face to face relations between labouring people at shopfloor, community and grass-roots level
- stimulating an international network model for labour organisation based upon self-empowerment, decentralisation, horizontal and democratic relations
- replacing an 'aid model' (one way flows of money and material from the 'rich, powerful, free' unions, workers or others) with a 'solidarity model' (two way or multi-directional flows of political support, information and ideas)
- refocusing priorities towards direct activity, visits and support from the grassroots level and away from verbal declarations, conferences and appeals
- while recognising internationalism to be essential to workers, refusing to see organised labour as 'the privileged bearer of internationalism' and thereby allowing workers to forge links with other forms of democratic internationalism
- linking solidarity abroad with action at home – combating racism, nationalism and discrimination of all kinds, locally
  (Waterman, 1998: 80)

New technologies, cheaper travel and the end of the Cold War are certainly altering the dynamics of international contact between workers (see Box 2.9 for the example of the Liverpool dock workers' dispute). Internationalism has continued to evolve in the 150 years since the publication of *The Communist Manifesto*, and it is certain to be a key facet of political life in the twenty-first century.

Box 2.9 Labour internationalism for the twenty-first century? The case of the Liverpool Dockers' Dispute, 1995–97

On 25 September 1995, 22 workers employed by the Torside docks company in Liverpool, England, were instructed to work overtime to get a ship ready for sailing, and were told that normal overtime arrangements would not apply. While they waited for their shop stewards to come and discuss the matter, five of these workers were sacked for disobeying instructions. By the next morning the Managing Director of Torside had sacked the whole work force and they promptly set up a picket line which other dockers refused to cross (see Figure 2.9.1). By 28 September 1995 Torside and the Mersey Docks and Harbour Company had sacked about 500 workers and begun to recruit an 'alternative' work force.

So began a bitter dispute which lasted for more than two years. This dispute broke new ground in the way in which rank and file workers organised international action with other dock workers across the world. The Liverpool

Box 2.9 *continued*

Figure 2.9.1 Liverpool Dockers on strike

dockers had a fourfold strategy in their international work. First, they sent pickets to distant ports to spread their message and to find common ground with other workers who were facing casualisation and falling living standards. Following many of these encounters, workers then organised unofficial industrial action to target ships bound for Liverpool (and the transatlantic shipping lines ACL, CAST and CanMar were particularly badly hit by American, Canadian and Swedish workers). Second, they set up an Internet site which was regularly updated with news on the dispute and the progress of solidarity support. Third, the dockers organised two international conferences (on 19–23 February and 31 August–1 September 1996) which were attended by rank and file trade union activists from at least 18 countries. Fourth, these conferences were the key to organising days of international solidarity action (on 30 September 1996 and 20 January 1997) when Swedish, Danish, French, Canadian, American and Australian dock workers (among others) took industrial action to target shipping companies which dealt with Liverpool. Although the International Transport Workers' Federation (ITF) (one of the International Trade Secretariats, see Box 2.8) supported the dockers, their activity and that of the Transport and General Workers' Union was limited by UK trade union law.

International activity was not limited to the dock workers themselves, and the wives, partners, daughters and mothers of the dockers set up *Women of the Waterfront* as an additional part of the campaign. These women took part in picketing, organising and fund-raising, and sent delegations to represent the dispute to women's groups and unions in different towns, cities and nations.

Box 2.9 *continued*

As this resolution demonstrates, the Liverpool dockers seized the political advantage of the Internet and all it offers for international communication between workers when fighting disputes:

RESOLUTION ON DOCKER INTERNATIONAL LABOUR COMMUNICATION

Whereas the need to link up dock and maritime workers world-wide is critical to exchange information and link our common struggles and,
Whereas the establishment of Labournet and the Dockers' World-Wide Web pages have been important in breaking the media blockade of information about the Liverpool dockers' struggle and,
Whereas, dock and maritime workers world-wide must develop open and broad communication links among all their fellow workers internationally,
Therefore, be it RESOLVED: "That this conference endorses the use of Labournet/Labornet and calls on all delegates and their affiliates and all maritime workers in the ITF to link up on the Internet and provide information on their issues and struggles on this international communication network and on Labournet/Labornet."

(passed unanimously at the International Dockworkers' Conference, Liverpool: www.labournet.org)

Sources: Dockers' World Wide Web Pages (www.labournet.org) and Lavalette and Kennedy, 1996. For additional material on the labour movement, internationalism and the use of the Internet, see Lee, 1997

Geographies of marxism: from the 'classical' to the 'western' tradition

In the years before and after the Second World War the marxist tradition began to develop beyond the preoccupations of those thinkers and activists we have described as 'classical' marxists. As socialism did not spread to the advanced economies, as Stalin secured an authoritarian grip upon the Russian people, and as the workers of the world were engaged in the mutual destruction of two world wars, some marxist theoreticians began to question the old certainties of the tradition. In particular, they moved away from practical involvement in the day-to-day organisations of the working class, shifting the focus of marxism from political-economy and the agency of the working class towards an interest in philosophy and the study of culture. This body of marxist thought has been called 'western' and, as Anderson argues, it is a tradition born of a different generation of thinkers, in different circumstances, living in different parts of the world:

In their [western marxist] hands, Marxism became a type of theory in certain critical respects quite different from anything that had preceded it. In particular, the characteristic themes and concerns of the whole ensemble of theorists who came to political maturity before the First World War were drastically displaced, in a shift that was at once generational and geographical. (Anderson, 1979: 25)

As indicated in Box 2.10 and Figure 2.4, the key protagonists in this set of debates, apart from Lukács and Goldmann, tended to originate and live in Germany, France and Italy (in contrast to the classical marxists who had been concentrated in Central and Eastern Europe). At a time when Stalin had asserted a firm ideological command over marxist scholarship and thought in the Soviet empire, the freedom to experiment with marxist ideas and concepts was found only in the West (although it was further limited by the growth of fascism during the 1930s). In these difficult times, 'western' marxism came into being in the places where thinkers had the ideological space to develop. But even in non-fascist western economies, the key enquiries of classical marxism (political-economy and working class agency) were dominated by Stalinist Communist Party intellectuals, leaving space only at the margins of marxism. In this context, original developments in marxist thought were made in matters of philosophical method, culture and aesthetics – beyond the gaze and control of the 'official' (Stalinist) marxist tradition (Anderson, 1979; Merquior, 1986; Gottlieb, 1989).

Box 2.10 The western marxists

	Dates	Place of birth	Key writings
György Lukács	1885–1971	Budapest, Hungary	*History and Class Consciousness* (1923); *Blum Theses* (1928); *The Specific Nature of the Aesthetic* (1962); *Towards an Ontology of Social Being* (1971)
Karl Korsch	1886–1961	Tostedt, Germany	*Marxism and Philosophy* (1923); *Karl Marx* (1938)
Antonio Gramsci	1891–1937	Ales, Sardinia	*Political Writings* (1910–26); *Prison Notebooks* (1929–35)
Walter Benjamin	1892–1940	Berlin, Germany	*One-Way Street and Other Writings* (1971); *Illuminations* (1973); *Understanding Brecht* (1977)
Max Horkheimer	1895–1973	Stuttgart, Germany	*Eclipse of Reason* (1947); *Dialectic of Enlightenment* (1947 with Adorno); *Critical Theory* (1968)
G. Della Volpe	1897–1968	Imola, Italy	*Logic as a Positive Science* (1950); *Rousseau and Marx* (1964)

Box 2.10 *continued*

Herbert Marcuse	1898–1979	Berlin, Germany	*Reason and Revolution (1941); Eros and Civilisation (1955); One Dimensional Man (1964); The Aesthetic Dimension (1978)*
Henri Lefebvre	1901–1991	Hagetmau, France	*Dialectical Materialism (1938); Critique of Everyday Life (1947); The Urban Revolution (1970); The Production of Space (1974); The Survival of Capitalism (1976)*
Theodor Adorno	1903–1969	Frankfurt, Germany	*Philosophy of Modern Music (1949); Minima Moralia (1951); Negative Dialectics (1966)*
Jean-Paul Sartre	1905–1980	Paris, France	*The Transcendence of the Ego (1936); Being and Nothingness (1943); Critique of Dialectical Reason (1960); Between Existentialism and Marxism (1972)*
Lucien Goldmann	1913–1970	Bucharest, Romania	*The Human Sciences and Philosophy (1952); Towards a Sociology of the Novel (1964)*
Louis Althusser	1918–1990	Birmandreis, Algeria	*For Marx (1965); Reading 'Capital' (1970, with Etienne Balibar); Lenin and Philosophy and Other Essays (1971)*
Lucio Colletti	1924–	Rome, Italy	*From Rousseau to Lenin (1969); Marxism and Hegel (1969)*

Sources: Anderson, 1979: 25–26; Bottomore et al., 1983

The majority of the thinkers associated with western marxism came to political maturity during the dark days of Stalinism, fascism and war, and it is hardly surprising that in this context they were pessimistic about the possibilities for positive change. Only Lukács and Gramsci played any major role in the class struggle during their lifetimes (these thinkers bridged the gap between classical and western marxism), and the majority of these theoreticians spent their days in relatively isolated scholarship and debate, often in universities. As

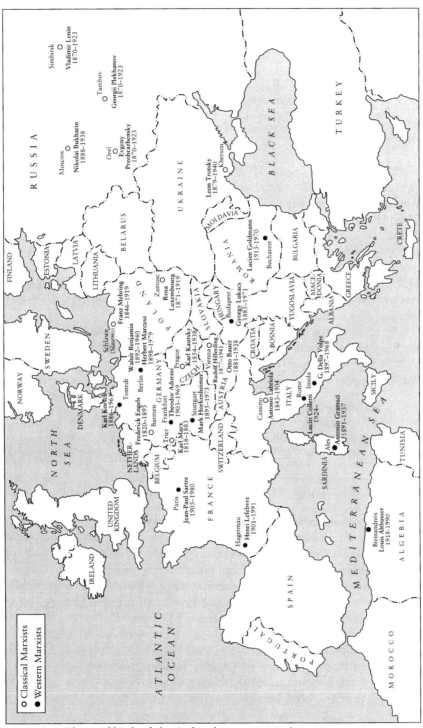

Figure 2.4 Places of birth of classical and western marxists

Anderson explains, the development of fascism and Stalinism shattered the classical marxist tradition by breaking the connection between theory and practice:

> Fascism and Stalinism ... combined to scatter and destroy the potential bearers for an indigenous Marxist theory united to the mass practice of the western proletariat ... Henceforward, it [Marxism] was to speak its own enciphered language, at an increasingly remote distance from the class whose fortunes it formally sought to serve or articulate. (Anderson, 1979: 32)

Divorced from the class struggle in a manner which would have been inconceivable to the previous generation of marxists such as Luxemburg, Lenin and Trotsky, this new generation of marxists had very different concerns. Many of the German scholars listed in Box 2.10 were associated with the *Frankfurt Institute of Social Research* which was founded in 1923 to promote marxist studies in a university context (although at the start of Nazi rule the Institute moved to Colombia University in the United States during 1933). The work produced by this school of thinkers became known as *critical theory*, and, as befits their distance from a mass readership, much of their writing is composed of very complex and difficult language.

During the twentieth century, then, marxism became a matter of academic concern, and Horkheimer's notion of critical theory reflected this concern with methods of knowledge creation. David Held describes critical theory as having a threefold approach:

> First, there was the idea of a critique of ideology which he took to be similar in structure to Marx's critique of capitalist commodity production and exchange. Second, there was a stress on the necessity of reintegrating disciplines through interdisciplinary research. Third, there was an emphasis on the central role of praxis in the ultimate verification of theories: The claim of critique to be the 'potential critical self-awareness of society' had to be upheld in particular. (Held, in Bottomore *et al.*, 1983: 214)

As this quotation demonstrates, western marxists often incorporated the ideas of Marx with the work of other social theorists and with other methods of social research. As a result, this tradition of marxism is very diverse and eclectic as the ideas of scholars such as Hegel, Lacan, Kant, Freud, Simmel, Spinoza and Weber (to name but a few) were combined with insights from Marx. Moreover, this diversity was reinforced by the lack of international contact between protagonists in the western marxist tradition. Scholars tended to be confined to their own university setting and it was only when their ideas were taken up in popular struggle (as happened to Sartre, Marcuse and Althusser during the 1960s and 1970s) that these thinkers became more generally known. Yet despite the difficulties that they faced, the western marxists did produce major new bodies of work in the realm of theory and method, art, aesthetics and culture, which all advanced the marxist heritage in new and exciting directions:

> Within its newly constricted parameters, the brilliance and fertility
> of this tradition were by any standards remarkable. Not merely did
> Marxist philosophy achieve a general plateau of sophistication far beyond
> its median levels of the past; but the major exponents of Western Marxism
> also pioneered studies of cultural processes – in the higher ranges of the
> superstructures – as if in glittering compensation for their neglect of the
> structures and infrastructures of politics and economics. (Anderson, 1983: 17)

Much of this literature has yet to be reflected within the geographical discipline, though for some connections and an overview, see Gregory (1978, 1994). As we have seen, the key developments in marxist geography have drawn inspiration from the classical rather than the western tradition of thought. The work of Lefebvre is, however, a major exception to this and he was the only marxist scholar in the western tradition who focused on geographical concerns. Lefebvre produced a huge volume of work over his long lifetime and from 1968 onwards he produced seven books which explored urbanism and spatiality in more detail (for a summary of Lefebvre's life and writings, see Box 2.11). This work culminated in *The Production of Space* which is rich with geographical and political insights about the spatiality of social life, production, consumption and resistance. Although it was written in 1974, *The Production of Space* was not translated into English until 1991, and it is a very complex and difficult book (not least because it contains an enormous number of references and debates to other bodies of writing). Chapter 1 of the volume opens with the words:

> Not so many years ago, the word 'space' had a strictly geometrical meaning:
> the idea it evoked was simply that of an empty area ... to speak of 'social
> space' ... would have sounded strange. (Lefebvre, [1974] 1991: 1)

But then a few pages later, Lefebvre introduces the major focus of the rest of the book:

> Few people today would reject the idea that capital and capitalism 'influence'
> practical matters relating to space, from the construction of buildings to the
> distribution of investments and the worldwide division of labour. (Lefebvre,
> [1974] 1991: 9–10)

Box 2.11 Henri Lefebvre

Born in 1901, Henri Lefebvre had a remarkably productive life, writing almost 70 books by the time of his death in 1991. Lefebvre studied at the Sorbonne, Paris, during the 1920s and was attracted to the avant-garde movements of the time, taking part in a philosophical discussion group. Strongly influenced by the successful Russian revolution, Lefebvre joined the French Communist Party in 1928 before becoming a school teacher. During the Second World War he had to flee Paris and he joined the Resistance, first in Marseilles and then in the Campan Valley in the Pyrenees. This afforded Lefebvre the opportunity to

Box 2.11 *continued*

undertake detailed sociological studies of peasant life which were eventually published in 1963 (as *La Vallée de Campan*).

By this time Lefebvre had already made a serious contribution to the intellectual life of the French Communist Party. His *Le Matérialisme Dialectique*, published in 1939, had been particularly well received in contrast to his earlier work on alienation (*La Conscience Mystifiée*, written with Herbert Guterman). From 1945 he had a research position with the Centre National de Recherche Scientifique in Paris, but his break with the Communist Party came with the publication of the Khrushchev Report in 1956. This report detailed some of the atrocities of the Stalinist regime and made it impossible for Lefebvre to remain a member of the Party – although it is significant that he was expelled rather than resigning, perhaps reflecting his deep commitment to the organisation he had served for 30 years.

This freedom from 'official' marxism then allowed Lefebvre to enlarge upon his ideas while developing new themes in his writing. As a university professor (in Strasbourg from 1961 to 1965 and in Nanterre, Paris, from 1965 to 1973), he challenged positivism, Althusserian structuralism and the early work of post-structuralists as well as the existentialism of Sartre and Merleau-Ponty. Following his involvement in the protests of 1968 (Nanterre University being the place where student protest began), Lefebvre delved deeper into questions of urbanism, space and political protest. As David Harvey explains:

> Increasingly during the 1960s, and particularly through the events of 1968, Lefebvre came to recognise the significance of urban conditions of daily life (as opposed to narrow concentration on work-place politics) as central in the evolution of revolutionary sentiments and politics. The significance of the outbreak in Nanterre – a suburban university close to the impoverished shanty-towns of the periphery – and the subsequent geography of street actions in Paris itself, alerted him to the way in which these kinds of struggle unfolded in a distinctively urban space. (Harvey, 1991: 430)

Lefebvre's analysis of spatiality and everyday life were his greatest legacy to the marxist tradition. And as Ed Soja explains in rather purple prose, Lefebvre's interest in space was born of his lifetime experiences and his particular reading of marxist and radical thought:

> His was always a deeply peripheral consciousness, inherently heretical and contracentric, a consciousness shaped in the regions of resistance beyond the established centres of power yet peculiarly able to comprehend the innermost workings of the power centres, to know their perils and possibilities, to dwell within them with the critical ambidexterity of the resident alien, the insider who always purposefully remains outside. (Soja, 1991: 257)

Source: Harvey, 1991

In *The Production of Space*, Lefebvre outlined a marxist approach to understanding the connections between society and space. In particular, he developed a conceptual triad to highlight the different ways and processes through which space is constantly produced. Contrasting perceived, conceived and lived space, Lefebvre distinguished between the spatiality of each, focusing upon spatial practices, representations of space, and representational spaces (or spaces of representation) in the following manner:

1. *Spatial practice*, which embraces production and reproduction, and the particular locations and spatial sets characteristic of each social formation. Spatial practice ensures continuity and some degree of cohesion. In terms of social space, and of each member of a given society's relationship to that space, this cohesion implies a guaranteed level of competence and a specific level of performance.
    Sites, circuits and environments through which social life is produced and reproduced.

2. *Representations of space*, which are tied to the relations of production and to the 'order' which those relations impose, and hence to knowledge, to signs, to codes, and to 'frontal' relations.
    (The space of scientists, planners, urbanists, technocratic subdividers and social engineers.)
    The way in which the power, knowledge and spatiality of the powerful is inscribed in space.

3. *Representational spaces/Spaces of representation*, embodying complex symbolisms, sometimes coded, sometimes not, linked to the clandestine or underground side of social life, as also to art (which may come eventually to be defined less as a code of space than as a code of representational space).
    Counter-hegemonic spaces which allow challenges to the dominant order.

(Lefebvre, [1974] 1991: 33 (38); sentences in italics adapted from Gregory, 1994: 403)

In a capitalist society, these processes tend to produce and reproduce (although not without contradictions between them at different times and in different places) an 'abstract' space:

> Capitalism and neo-capitalism have produced abstract space, which includes the 'world of commodities', its 'logic' and its worldwide strategies, as well as the power of money and that of the political state. This space is founded on the vast network of banks, business centres and major productive entities, as also on motorways, airports and information lattices. (Lefebvre, [1974] 1991: 53)

Lefebvre argued that in capitalist society, space becomes part of the mode of production:

> Space as a whole enters into the modernized mode of capitalist production: it is utilized to produce surplus value. The ground, the underground, the air, and even light enter into both the productive forces and their products. The urban fabric, with its multiple networks of communication and exchange, is part of the means of production. The city and its various installations (ports, tube trains, etc.) are part of capital. (Lefebvre, 1979: 287)

Furthermore, the ideological dominance of the ruling class is argued to ensure that this form of space is produced largely through consent – without the need for generalised violence or force against those without power – as everyday life is increasingly penetrated by capitalist market relations:

> Abstract space, the space of the bourgeoisie and of capitalism, bound up as it is with exchange (of goods and commodities, as of written and spoken words, etc.) depends on consensus more than any space before it.
> (Lefebvre, [1974] 1991: 57)

For Lefebvre, as for the classical marxists who explored the uneven development of capitalism, the end of abstract space depended upon the class struggle – broadly defined to include oppositional movements beyond purely working class organisations. In *The Production of Space*, Lefebvre argued that active class struggle can differentiate and rupture the abstract nature of capitalist space by realising an alternative spatial order and a new form of space:

> Today, more than ever, the class struggle is inscribed in space. Indeed, it is that struggle alone which prevents abstract space from taking over the whole planet and papering over all differences. Only the class struggle has the capacity to differentiate, to generate differences which are not intrinsic to economic growth ... that is to say, differences which are neither induced by nor acceptable to that growth. (Lefebvre, [1974] 1991: 55)

Lefebvre's expectations of this process of resistance, which stems from the third part of his conceptual triad (lived space in which people can realise their own spaces of representation or representational spaces), are most clearly illustrated in his critique of Stalinism. By positing that revolutions can only succeed by creating a new space, Lefebvre described the way in which Stalinist Russia had come to reflect capitalism, a place of abstract space in which superstructural changes had not been founded on new forms of social organisation:

> A revolution that does not produce a new space has not realized its full potential; indeed it has failed in that it has not changed life itself, but has merely changed ideological superstructures, institutions or political apparatuses. A social transformation, to be truly revolutionary in character, must manifest a creative capacity in its effects on daily life, on language, and on space – though its impact need not occur at the same rate, or with equal force, in each of these areas. (Lefebvre, [1974] 1991: 54)

As this quotation demonstrates, Lefebvre retained a close interest in what he called 'everyday life' and it was from the daily interactions between people that he saw hope for the future. As Soja (1991: 258) has written: 'For Lefebvre, the lived spaces were passionate, "hot", teeming with sensual intimacies'. Such analysis bears testament to Lefebvre's continued belief that another world could be made and sustained; and in this regard, as in his geographical interests, Lefebvre was very different from the majority of his marxist contemporaries in the universities of western Europe.

While Lefebvre represents the major western marxist influence on the discipline of geography (his work also featuring in the writings of David Harvey and Ed Soja), the ideas of Louis Althusser have also been translated into urban studies by Manuel Castells.[2] As Castells wrote in the preface *to The Urban Question: A marxist approach*: 'I have proposed an adaptation of Marxist concepts to the urban sphere, using in particular the reading of Marx given by the French philosopher Louis Althusser' (Castells, [1972] 1977: ix). Mirroring Althusser's formulation of an epistemological break in the work of Marx (between his youthful ideas in the Hegelian tradition and his more mature constructions of dialectical materialism as a science of society), Castells argued that urban sociology required an epistemological break away from idealism and empiricism to science.[3] Castells thus proposed that the science of historical materialism (as articulated by Althusser) could explain the production of urban space and the structure and function of cities. In particular, he suggested that the urban should be understood as part of the total system of capitalism, and, in Althusserian language, the city was argued to be 'over determined' by a series of different practices and structures, including the economic (production, consumption and exchange), the political (state administration) and the ideological (symbolic processes). As part of this whole, however, the city was argued to perform a very specific function for capitalism – the reproduction of labour power – which is the only social process fundamental to capitalism which takes place purely at the scale of the urban (unlike production or exchange, or political administration which reach beyond the borders of cities). Castells argued that the processes of consumption involved in the reproduction of labour power are concentrated in urban space, particularly as the state takes on the role of service provider (in the realm of housing, health care and education, for example). In this regard, then, Castells was able to suggest that the urban was a unit of collective consumption, and, as such, the urban became politicised. Without suggesting that class conflict or social movements would necessarily arise from the politics of collective consumption, Castells argued that urban politics was over-determined, being causally related to a range of different structures and processes in operation:

> The permanent and ever extending intervention of the state apparatus in the area of the processes and units of consumption makes it the real source of order in everyday life. This intervention of the state apparatus, which we call urban planning, in the broad sense, involves an almost immediate politicization of the whole urban problematic … However, the politicization thus established is not necessarily a source of conflict or change, for it may also be a mechanism of integration and participation: everything depends on the articulation of the contradictions and practices. (Castells, [1972] 1977: 463)

As Castells further developed his ideas and research he moved away from Althusserian marxism, and, in common with the social sciences more generally, the geographical legacy of Althusser is now felt in post-structural theory and research. Purely Althusserian interpretations of Marx no longer have great currency, but for our purposes this interlude demonstrates the diversity of marxist scholarship and research. It is interesting that although both Lefebvre and Castells had similar goals, each seeking to unpack the relationships between capitalist society and the production of space using marxist ideas, each produced very different perspectives on space. While Castells argued that the city was a product of the complex social structures and practices of capitalist society, in which human beings had no conscious or active role, Lefebvre preferred to see the city as a dynamic site of human activity, creativity and conscious organisation. In this sense, Lefebvre retained a closer connection to the classical marxist tradition whose thinkers suggested that capitalism could be overthrown from within, by self-conscious, working class organisation.

This brief discussion of western marxism would seem to suggest that it is still the classical tradition which has borne greatest fruit in geographical thinking – particularly when the ideas of Lefebvre are found to differ in many respects from those of other theorists in the western marxist tradition (his faith in active political agency connecting him to those earlier thinkers). Yet the recent renewal of cultural geography bears testament to another body of twentieth century marxism, which has been broadly labelled as 'British'. Although not usually included in the corpus of western marxism, British marxists such as Edward Thompson, Eric Hobsbawm and Raymond Williams have played a significant role in the evolution of marxist scholarship and research, particularly in historiography and cultural studies.[4] Moreover, as human geographers have developed new interests in cultural studies during the 1980s and 1990s, Williams's concept of cultural materialism has played a key role in disciplinary debate (see Box 2.12). In this regard then, marxism (in all its diversity) remains important to understanding the geography of capitalism and the geography of contemporary politics and culture.

Box 2.12 Raymond Williams and cultural materialism

Raymond Williams was born in Pandy, on the Welsh borders, of strong working class roots. This background left him with a profound ambivalence about his academic position at Cambridge University where he became a Fellow in 1961 and Professor of Drama in 1974. Although he was active in the Communist Party for a short time before the Second World War, Williams remained outside organised politics for the rest of his life, yet his writings (both fictional and academic) were profoundly political, and he retained a strong interest in Marx's ideas.

Williams is best known for developing a materialist approach to culture (in *Culture and Society* (1958) and *Marxism and Literature* (1977)), which, in contrast to structural marxism, does not 'read' culture from the economic base of society. While he argued that culture is produced, and as such, is rooted in the nature of society and the relations between individuals, Williams suggested that the relationship between society and culture is not unidirectional or straightforwardly causal:

Box 2.12 *continued*

> From castles and palaces and churches to prisons and workhouses and schools; from weapons of war to a controlled press: any ruling class, in variable ways though always materially, produces a social and political order. These are never superstructural activities ... The complexity of this process is especially remarkable in advanced capitalist societies where it is wholly beside the point to isolate 'production' and 'industry' from the comparably material production of 'defence', 'law and order', 'welfare', 'entertainment', and 'public opinion'. (Williams, 1977: 93)

Williams overturned certain deterministic readings of the relationship between base and superstructure (see Box 2.5) to argue that the economy merely 'limits' the possibilities of cultural expression and political organisation rather than 'determining' them in any mechanical sense. As Peter Jackson has argued in relation to cultural geography,

> Williams offers a view of determination that is thoroughly appropriate to a reconstituted cultural geography. Rather than seeing determination as something that takes place in relation to a static mode of production, he adopts a more active, conscious view of historical experience, recognizing multiple forces of determination, structured in particular historical situations. (Jackson, 1989: 35–36)

Williams placed cultural processes in their social and historical setting, arguing that culture is a 'signifying system' and as such, culture is part of:

> ... any economic system, any political system, any generational system and, most generally ... any social system. Yet it is also in practice distinguishable as a system in itself: as a language, most evidently; as a system of thought or of consciousness, or, to use that difficult alternative term, an ideology; and again as a body of specifically signifying works of art and thought. Moreover, all these exist not only as institutions and works, and not only as systems, but necessarily as active practices and states of mind. (Williams, 1981: 207-208)

As this quotation attests, Williams understood culture to be a separate but integrated part of everyday life, and as such, it is argued to be impossible to make sense of the economy, politics or society without attention to culture. In his hands, marxism is thus used as a way to integrate disciplinary concerns and understanding to make sense of the whole – in a way that is reminiscent of Horkheimer's model of critical theory. Moreover, the legacy of Williams's cultural materialism is witnessed in contemporary cultural geography (Jackson, 1989; Cosgrove, 1984; Daniels, 1989) and in the 'cultural turn' which has shaped human geography more generally (for examples from economic geography, see the first section of Lee and Wills, 1997).

Sources: Jackson, 1989; Williams, 1977

Yet there is more to the encounter between marxism and geography than the development of marxist geography. The geographical engagement with Marx and Lefebvre (in particular) has prompted a rethinking of marxism, and social theory more generally, to include a spatial dimension. While history and time were afforded priority over space in twentieth-century thinking, the development of marxist geography has contributed to the growing realisation that 'Marxism itself had to be critically restructured to incorporate a salient and central spatial dimension' (Soja, 1989: 335; see also Massey, 1992). In his attempt to reconfigure historical materialism as historical–geographical materialism, David Harvey's work has been critically important in the spatialisation of marxism. In *The Condition of Postmodernity* (1989), Harvey outlines what this new approach means for marxism, and the value of marxist theory in understanding political-economy alongside culture, politics and subjectivity. He cites four particularly important developments in the spatialisation of marxism:

1. The treatment of difference and 'otherness' not as something to be added on to more fundamental Marxist categories (like class and productive forces), but as something that should be omni-present from the very beginning in any attempt to grasp the dialectics of social change. The importance of recuperating such aspects of social organisation as race, gender, religion, within the overall frame of historical materialist enquiry (with its emphasis upon the power of money and capital circulation) and class politics (with its emphasis upon the unity of the emancipatory struggle) cannot be overestimated.
    [*i.e. social relations other than class are important to explaining both social life and social change*]

2. A recognition that the production of images and of discourses is an important facet of activity that has to be analysed as part and parcel of the reproduction and transformation of any symbolic order. Aesthetic and cultural practices matter, and the conditions of their production deserve the closest attention.
    [*i.e. culture is part of the material makeup of society, the way we think about it and live within it – Williams's cultural materialism*]

3. A recognition that the dimensions of space and time matter, and that there are real geographies of social action, real as well as metaphorical territories and spaces of power that become vital as organising forces in the geopolitics of capitalism, and at the same time as they are the sites of innumerable differences and othernesses that have to be understood both in their own right and within the overall logic of capitalist development. Historical materialism is finally beginning to take its geography seriously.
    [*i.e. marxism can be strengthened by taking geography seriously, space is fundamental to the nature of the world, its production, consumption, political activity and experience*]

4. Historical–geographical materialism is an open-ended and dialectical mode of enquiry rather than a closed and fixed body of understandings. Meta-theory is not a statement of total truth but an attempt to come to terms with the historical and geographical truths that characterise capitalism both in general as well as in its present phase.
    [*i.e. marxism is not a source of absolute truth but rather a method, a means of starting to understand current changes in the world*]

(Harvey, 1989: 355 [text in brackets added for explanation]).

Since the 1970s there has been a productive encounter between marxism and geography which has marxified geography and spatialised marxism to the benefit of both, but the future directions of this exchange are hard to predict.

The future of marxism?

Over the course of the twentieth century, the marxist tradition came to reflect the changing history and geography of each period. The unity of theory and practice of the early, optimistic years of the century collapsed under the weight of Stalinism, fascism and war. Many marxist scholars were shaken by the destructive power of Stalinism and were able to survive only by retreating into the universities to produce complex treatises on matters outside the classical marxist tradition. While this new generation of western marxists did reflect societal anxieties about the stabilisation of capitalism, and the importance of ecology, sexuality, ideology, bureaucracy and alienation, they did so largely without any commitment to revolutionary organisation. Even the temporary renewal of interest in marxist ideas during the 1960s and 1970s did not stop this shift away from any faith in the agency of the working class to change the world for the better. Indeed, it can be argued that the experiences of the 1960s and 1970s gave birth to a wave of new social movements based on race, gender and sexuality, rather than class – and in this situation, marxism has faded further from view, as new forms of protest, based on new social agents, with new social agendas, came into their own (as discussed in Chapters 3, 4 and 5).

In many ways, this decline is rooted in the history of working class struggle. As the Russian revolution failed to spread internationally, quickly degenerating into Stalinism, the dreams of the early generation of marxists faded from view. The twentieth century has witnessed non-working class revolutions and working class struggles which have been contained by the capitalist system (although the welfare gains that workers have made in advanced capitalist economies have come at a price to employers – as demonstrated in reverse by the cuts that have been enacted since the 1970s and 1980s). Moreover, even though the working class continues to expand globally, there is little evidence of a resurgence in revolutionary fervour, as David Harvey notes in his reading of *The Communist Manifesto*:

> The global proletariat is far larger than ever and the imperative for workers to unite is greater than ever. But the barriers to that unity are far more formidable than they were in the already complicated European context of 1848. The workforce is now far more geographically dispersed, culturally heterogeneous, ethnically and religiously diverse, racially stratified, and linguistically fragmented. The effect is to radically differentiate both the modes of resistance to capitalism and the definitions of alternatives.
> (Harvey, 1998: 68)

While it is possible to challenge such a pessimistic view of the possibilities for international working class action (see Moody, 1997; Waterman, 1998; Wills,

1998), the experiences of the recent past tend to bear Harvey out. The marxist tradition has hit a low point, and although Marx's ideas and the classical marxist tradition have been 'liberated' by the collapse of Stalinism and the fall of the Berlin Wall, the damage that has been done will take a long time to heal. While there is now space for the reinterpretation of marxist ideas to meet the challenges of the contemporary world, there are few people engaged in this project in geography (see Gibson-Graham, 1996; Castree, 1999). Moreover, without some connection to those trying to change the world, marxism is likely to remain a rather abstract and scholarly affair, being divorced from those it purportedly seeks to embrace. For the future, the connections between theory and practice must take on central importance if any revival is to take place. For while marxism still has enormous relevance as a critique of capitalism and its social relations, it is a living tradition only so long as it speaks to the present, to the future, and to those who challenge the established order of things. The future of marxism (and both its practical and disciplinary geographical manifestations) thus depends on a new generation of thinkers and activists, grappling with new conditions in a fast-changing world.

Conclusions

The ideas of Marx have spawned a very rich and diverse tradition of thought and practice. After outlining the broad contours of Marx's ideas, this chapter has focused on the geographical insights and implications of the tradition. In particular, marxists have shown how capitalism is characterised by uneven development, and how this unevenness then shapes the further evolution of capitalism and associated political protest. Working class organisation has always been differentiated across space as well as time and marxists have sought to build solidarity between workers in different locations.

This chapter has highlighted the ways in which the marxist tradition has evolved over time as thinkers and activists find themselves in changing circumstances and in need of new understanding. After the Stalinisation of the Russian revolution and the failure of socialism to spread to the advanced capitalist nations, the focus of marxist intellectual development shifted westwards. The so-called western marxists moved away from an active engagement in working class organisation and from the study of political-economy to focus on questions of philosophy and culture. Based mainly in the universities, marxism developed as a form of critical theory in the years after the Second World War, its intellectuals producing works on method, art, aesthetics and cultural studies. As a part of this tradition of dissent, the work of Henri Lefebvre is particularly significant for geography as he applied marxist concepts to understand the production of space. As illustrated in this chapter, the productive encounter between marxism and geography has involved both the classical and western strands of the tradition. As a result, human geography has been immeasurably strengthened while marxists have also come to recognise the importance of space.

Notes:

1.   It is important to acknowledge that all the major revolutionary movements of the twentieth century have indeed taken place in countries on the margins of the capitalist world economy. The revolutions in China (1949), Cuba (1959) and Nicaragua (1979) all took place in largely peasant economies where capitalism was poorly developed. However, in contrast to the Russian revolution, the organised working class did not play the central role in determining the course of these revolutionary movements. Whereas Trotsky suggested that the working class were in a key position to lead revolutions to socialism in such peripheral states, experience suggests that other social agencies (particularly peasant guerilla movements led by intellectuals) can also fulfil this role. Subsequent analysis has thus suggested that these cases represent 'deflected permanent revolution' (see Cliff, 1986). In the contemporary world economy it is extremely difficult for states to survive after revolutionary upheaval (particularly as the Soviet Union is no longer able to support such regimes), making permanent revolution without international support more problematic than ever.

2.   While the work of Antonio Gramsci has not spawned a particular reading of the producion of space or a particular school of thinking within geographical thought, a number of his ideas have had important bearing upon the social sciences since the 1970s and 1980s. In particular, his conception of 'hegemony', described as the dominant web of beliefs and ideas which are fostered by the institutions and intellectuals of any society, has been widely adopted (for a geographical example, see Soja and Hadjimichalis, 1979). As capitalist society advanced, Gramsci documented the complex ways in which ideas became dominant, being actively reproduced in everyday life as 'common sense', and thus obfuscating the exploitative social relations of capitalism. Gramsci conceived the role of the revolutionary party as being critical in fostering a new layer of 'organic intellectuals' who could plant the seeds of a new hegemony, countering capitalism and its commonsense structures of feeling. Recently, Gramscian ideas have entered the geographical discipline via regulation theory which explores the way in which regimes of economic accumulation are dialectically interconnected to modes of social regulation (see Storper and Scott, 1986; Tickell and Peck, 1992, 1995).

3.   *The Urban Question: A marxist approach* is a complicated book and readers are advised to use the text by Saunders (1981) as a summary of Castells before tackling the work as a whole. Indeed, this summary has been partly drawn from Saunders who also provides a useful comparison between Castells and the work of Lefebvre.

4.   Anderson (1983) suggests that western marxism had run its course by the 1970s, largely due to the death of the protagonists, the explosions of mass protest in the late 1960s and 1970s which were not led by the left, but which presented new opportunities for connections to be made between theory and practice, and the collapse of the post-war boom which put the stability of capitalism into question. In this context, new bodies of marxist theory and research came to prominence, often looking at more traditional themes of political-economy, such as the labour process (Braverman, 1974), economic crisis (Mandel, 1978, 1980), regulation (Aglietta, 1979), the state (Poulantzas, 1978) and class (Wright, 1978). Moreover, Anderson points to the vitality of British-based marxist historiography as another key development in marxist theory and understanding. This new marxism, he suggests, has a different geography from the older models:

> The geographical pattern of Marxist theory has been profoundly altered in the past decade. Today the *predominant* centres of intellectual production seem to lie in the English-speaking world, rather than in Germanic or Latin Europe, as was the case in the inter-war and post-war periods respectively.... the traditionally most backward zones of the capitalist world, in Marxist culture, have suddenly become in many ways the most advanced. (Anderson, 1983: 24)

Moreover, Anderson goes on to suggest that this flowering of marxist work in the English-speaking world was mirrored by a collapse elsewhere:

> At the very time when Marxism as a critical theory has been in unprecedented ascent in the English-speaking world, it has undergone a precipitous descent in the Latin societies where it was most powerful and productive in the post-war period. (Anderson, 1983: 30)

References

AGLIETTA, M. (1979) *A Theory of Capitalist Regulation: the US experience.* London: New Left Books.

ANDERSON, P. (1979) *Considerations on Western Marxism.* London: Verso.

ANDERSON, P. (1983) *In the Tracks of Historical Materialism.* London: Verso.

BARAN, P. (1957) *The Political Economy of Growth.* New York: Monthly Review Press.

BERLIN, I. (1963) [1939] *Karl Marx: his life and environment.* London: Oxford University Press.

BERMAN, M. (1983) *All That is Solid Melts into Air: the experience of modernity.* London: Verso.

BLAUT, J.M. (1975) Imperialism: the Marxist theory and its evolution. *Antipode*, 7: 1–19.

BLAUT, J.M. (1994) *The Colonizer's Model of the World.* London: Guilford.

BOTTOMORE, T., HARRIS, L., KIERNAN, V.G. and MILIBAND, R. (1983) *A Dictionary of Marxist Thought.* Oxford: Blackwell.

BRAVERMAN, H. (1974) *Labour and Monopoly Capitalism: the degradation of work in the twentieth century.* New York: Monthly Review Press.

CALLINICOS, A. (1983) *The Revolutionary Ideas of Karl Marx.* London: Bookmarks.

CASTELLS, M. (1977) [1972] *The Urban Question: a marxist approach.* London: Edward Arnold.

CASTREE, N. (1999) Envisioning Capitalism: geography and the renewal of marxian political economy. *Transactions of the Institute of British Geographers*, 24: 137–58

CLIFF, T. (1986) *Deflected Permanent Revolution.* London: Bookmarks.

COSGROVE, D. (1984) *Social Formation and Symbolic Landscape.* London: Croom Helm.

DANIELS, S. (1989) Marxism, culture and the duplicity of landscape. In R. Peet and N. Thrift (eds), *New Models in Geography*, vol. 2. London: Unwin Hyman, 196–220.

DUNFORD, M. and PERRONS, D. (1983) *The Arena of Capital.* New York: St Martin's Press.

EMMANUEL, A. (1972) *Unequal Exchange: a study of the imperialism of trade.* London: New Left Books.

FRANK, A.G. (1969) *Capitalism and Underdevelopment in Latin America.* New York and London: Monthly Review Press.

GIBSON-GRAHAM, J.K. (1996) *The End of Capitalism (as we knew it).* Oxford: Blackwell.

GOTTLIEB, R.S. (ed) (1989) *An Anthology of Western Marxism.* Oxford: Oxford University Press.

GREGORY, D. (1978) *Ideology, Science and Human Geography.* London: Hutchinson.

GREGORY, D. (1994) *Geographical Imaginations.* Oxford: Blackwell.

HARVEY, D. (1975) The geography of capitalist accumulation: a reconstruction of the Marxian theory. *Antipode,* 7: 9–21.

HARVEY, D. (1982) *The Limits to Capital.* Oxford: Blackwell.

HARVEY, D. (1989) *The Condition of Postmodernity.* Oxford: Blackwell.

HARVEY, D. (1991) Afterword. In H. Lefebvre, *The Production of Space.* Oxford: Blackwell, 425–434.

HARVEY, D. (1998) The geography of class power. In L. Panitch and C. Leys (eds), *Socialist Register: the Communist Manifesto now.* London: Merlin, 49–74.

HEROD, A. (1995) The practice of labor solidarity and the geography of FDI. *Economic Geography,* 71: 341–363.

HOBSBAWM, E. (1998) Introduction. In K. Marx and F. Engels, *The Communist Manifesto: a modern edition.* London: Verso, 1–30.

HYMAN, R. (1983) *Marxism and the Sociology of Trade Unions.* London: Pluto

INTERNATIONAL LABOUR ORGANISATION (1920) International Congress of Workers in the Food and Drink Trades. *Studies and Reports,* series A.

JACKSON, P. (1989) *Maps of Meaning.* London: Routledge.

KORSCH, K. (1938) *Karl Marx.* London: Chapman and Hall.

LAPIDES, K. (1987) *Marx and Engels on the Trade Unions.* New York: Praeger.

LAVALETTE, M. and KENNEDY, J. (1996) *Solidarity on the Waterfront.* Liverpool: Liver Press.

LEE, E. (1997) *The Labour Movement and the Internet: the new internationalism.* London: Pluto.

LEE, R. and WILLS, J. (eds) (1997) *Geographies of Economies.* London: Arnold.

LEFEBVRE, H. (1979) Space: social product and use value. In J.W. Freiberg (ed.), *Critical Sociology: European perspectives.* New York: Irvington, 285–295.

LEFEBVRE, H. (1988) Toward a leftist cultural politics: remarks occasioned by the centenary of Marx's death. In C. Nelson and L. Grossberg (eds), *Marxism and the Interpretation of Culture.* London: Macmillan, 75–88.

LEFEBVRE, H. (1991) [1974] *The Production of Space.* Oxford: Blackwell.

LENIN, V.I. (1934) [1917] *Imperialism: the highest stage of capitalism.* 88London: Lawrence and Wishart.

LORWIN, L. (1929) *Labor and Internationalism.* New York: Macmillan.

LUXEMBURG, R. (1951) [1913] *The Accumulation of Capital.* London: Routledge and Kegan Paul.

LUXEMBURG, R. (1972) [1921] The Accumulation of Capital: an anti-critique. In R. Luxemburg and N. Bukharin, *Imperialism and the Accumulation of Capital.* London: Penguin, 47–140.

MACSHANE, D. (1996) *Global Business: global rights.* Fabian Pamphlet 575. London: Fabian Society.

MANDEL, E. (1978) *Late Capitalism.* London: Verso.

MANDEL, E. (1980) *Long Waves in Capitalist Development.* Cambridge: Cambridge University Press.

MARX, K. (1950) [1852] The Eighteenth Brumaire of Louis Bonaparte. In K. Marx and F. Engels, *Selected Correspondence,* Moscow: Foreign Language Publishing House, 221–311.

MARX, K (1950) [1871] General Rules of the International Working Men's Association. In K. Marx and F. Engels, *Selected Correspondence.* Moscow: Foreign Language Publishing House, 350–353.

MARX, K. (1961) [1867] *Capital: a critical analysis of capitalist production,* vol. I. Moscow: Foreign Language Publishing House.

MARX, K. (1973) [1857-58] *Grundrisse: foundations of the critique of political economy.* London: Penguin.

MARX, K. and ENGELS, F. (1970) [1846] *The German Ideology.* London: Lawrence and Wishart.

MARX, K. and ENGELS, F. (1971) [1848; German edition 1872] *Manifesto of the Communist Party.* Moscow: Progress Publishers.

MASSEY, D. (1992) Politics and space/time. *New Left Review,* **196**: 65–84.

MASSEY, D. (1994) [1984] *Spatial Divisions of Labour* (2nd edn). London: Macmillan.

MEHRING, F. (1936)[1918] *Karl Marx: the story of his life.* London: George Allen and Unwin.

MERQUIOR, J.G. (1986) *Paladin Movements and Ideas: Western Marxism.* London: Paladin.

MCLELLAN, D. (1973) *Karl Marx: his life and thought.* London: Macmillan.

MCLELLAN, D. (ed.) (1988) *Marxism: essential writings.* Oxford: Oxford University Press.

MILNER, S. (1990) *The Dilemmas of Internationalism: French syndicalism and the international labour movement, 1900–1914.* Oxford: Berg.

MOODY, K. (1997) *Workers in a Lean World: unions in the international economy.* London: Verso.

POULANTZAS, N. (1978) *State, Power, Socialism.* London: Verso.

SAUNDERS, P. (1981) *Social Theory and the Urban Question.* London: Unwin Hyman.

SLATER, D. (1977) Geography and underdevelopment. *Antipode,* **9**: 1–31.

SMITH, N. (1990) [1984] *Uneven Development.* Oxford: Blackwell.

SOJA, E. (1989) *Postmodern Geographies: the reassertion of space in critical social theory.* London: Verso.

SOJA, E. (1991) Henri Lefebvre 1901–1991. *Environment and Planning D: Society and Space,* **9**: 257–259.

SOJA, E. and Hadjimichalis, C. (1979) Between geographical materialism and spatial fetishism: some observations on the development of marxist spatial analysis. *Antipode,* **11**: 3–11.

STORPER, M. and SCOTT, A.J. (1986) Overview: production, work, territory. In A.J. Scott and M. Storper (eds), *Production, Work, Territory: the geographical anatomy of industrial capitalism.* Boston, MA: Allen and Unwin, 1–15.

THRIFT, N. and WILLIAMS, P. (eds) (1987) *Class and Space: the making of urban society.* London: Routledge and Kegan Paul.

TICKELL, A. and PECK, J. (1992) Accumulation, regulation and the geographies of post-Fordism: missing links in regulationist research. *Progress in Human Geography,* **16**: 190–218.

TICKELL, A. and PECK, J. (1995) Social regulation after Fordism: regulation theory, neo-liberalism and the global–local nexus. *Economy and Society,* **24**: 357–386.

TROTSKY, L. (1969) [1929] *The Permanent Revolution and Results and Prospects.* New York: Pathfinder.

TROTSKY, L. (1977) [1930] *The History of the Russian Revolution.* London: Pluto.

WATERMAN, P. (1998) *From Labour Internationalism to Global Solidarity.* London: Cassell.

WILLIAMS, R. (1977) *Marxism and Literature.* Oxford: Oxford University Press.

WILLIAMS, R. (1981) *Culture.* London: Fontana.

WILLS, J. (1998) Taking on the CosmoCorps: experiments in transnational labor organization. *Economic Geography,* **74**: 111–130.

WRIGHT, E.O. (1978) *Class, Crisis, State.* London: New Left Books.

3

Embodying Geography: Feminist Geographies of Gender

Feminism

Feminism is a political movement that seeks to overturn gender inequalities between men and women. According to Schneir, 'Feminism is one of the basic movements for human liberty' (Schneir, 1996: xi), seeking human liberty for women as well as men by challenging power relations that favour men and masculinity over and above women and femininity in different spheres of economic, social, political and cultural life. As Anne Phillips writes, 'Feminism is politics' (Phillips, 1998: 1). Feminism challenges and resists the gender roles and relations that position men and women in different and unequal ways in society. As such, feminism is concerned with the power relations that influence not only how individuals relate to each other, but how all spheres of life are gendered in particular ways. Feminism is a diverse political movement, varying over space and time. Reflecting its breadth, Griselda Pollock describes feminism as 'a political commitment to women and to changes that women desire for themselves and for the world' (Pollock, 1996: xv), while Lovenduski and Randall write that feminism spans 'all ideologies, activities, and policies whose goal it is to remove discrimination against women and to break down the male domination of society' (Lovenduski and Randall, 1993: 2).

Feminist politics are closely tied to feminist theory, and both are broad, diverse and contested fields. Feminist theory is concerned with analysing and explaining as well as changing gendered power relations. In its many different forms, feminist theory – or, perhaps more accurately, feminist *theories* – has inspired critical work across the humanities, social sciences and natural sciences that seeks to disrupt the gender imbalance of power that exists both within and beyond the academy. Although this work is diverse, reflecting different visions of feminism, politics and feminist politics, its concern with resisting and overturning gender imbalances remains constant. As Rosemarie Tong explains:

> What continues to fascinate me … is the way in which these partial and
> provisional answers intersect, joining together both to lament the ways in
> which women have been oppressed, repressed, and suppressed and to
> celebrate the ways in which so many women have 'beaten the system', taken
> charge of their own destinies, and encouraged each other to live, love, laugh,
> and be happy *as women*. (Tong, 1989: 1–2)

Since the late 1970s, and particularly over the course of the 1980s and
1990s, the work of feminist geographers has explored the connections between
gender and geography and has challenged gender inequalities in both geographical discourse – knowledge about the world – and disciplinary geography
(see McDowell and Sharp, 1997, 1999; Jones *et al.*, 1997; McDowell, 1999;
WGSG, 1997, for helpful overviews of feminist geographical work; and see
papers published in *Gender, Place and Culture*, to show the diversity of feminist geographies). Feminist geographies address 'the various ways in which
genders and geographies are mutually constituted' (Pratt, 1994: 94) by studying the ways in which *space is gendered* and also the ways in which *gender is
spatial*. But feminist geographies also reveal and resist the ways in which geography as a *discipline* is gendered, as shown by the disproportionate employment and seniority of men and women and by the politics, content and
methods of geographical research and teaching. The work of feminist geographers seeks to change gender inequalities both in the world and in the discipline
of geography.

While feminist geographies are diverse, they share several common themes
(as outlined by Pratt, 1994). One important site of resistance is against sexism
within the discipline, departments and other institutions of geography. Resisting the status of detached and disembodied analyses, feminist geographies
often share a commitment to situating the production of knowledge in more
embodied and contextualised ways. Finally, just as feminist theory and politics
transcend disciplinary boundaries, they also transcend the subdisciplinary
boundaries within geography, inspiring a wide range of connections within and
between economic, social, cultural, historical, political and urban fields of geographical interest. Feminist critiques of science have also begun to inspire critical studies of physical geography (see the themed issue of *Area* 30 on 'women,
gender, feminisms: visiting physical geography', 1998, in particular the papers
by Dumayne-Peaty and Wellens and by McEwan).

This chapter will begin by introducing different ideas about sex and gender.
It will then turn to focus on feminist politics in Britain and the United States.
Here, the so-called 'first-wave' feminist politics of the late nineteenth and early
twentieth centuries will be discussed alongside 'second-wave' feminist politics
from the 1960s and 1970s, which inspired feminist geographical work from
the late 1970s. Then some of the ways in which geographers have studied gender will be discussed, spanning liberal, socialist and poststructuralist traditions. Finally, the links between gender and geography as a discipline will
be explored, considering feminist histories of geography and the gendered
production of geographical knowledge.

Sex and gender

Gender is one part of everyone's identity, influencing how we think about ourselves, about other people, and about our relationships with other people. But just as identities extend far beyond the level of an individual, gender is also a social relation that positions men and women differently in different spheres of life. Both men and women have gender identities, which are often thought of in terms of masculinity and femininity. But 'gender' is usually distinguished from 'sex'. While sex is usually interpreted in terms of the biological or anatomical difference between men and women, gender is seen as a social construction that varies over space and time. According to McDowell, gender represents

> differences between women's and men's attitudes, behaviour and
> opportunities that depend upon socially constructed views of femininity and
> masculinity. The term gender is preferred to that of sex, which is restricted to
> the anatomical distinction between the sexes rather than social differences.
> (McDowell, 1986: 170)

In other words, to be born male or female does not imply a masculine or a feminine gender identity. Rather, ideas about masculinity and femininity are socially constructed. As Figure 3.1 shows, anatomical difference alone cannot

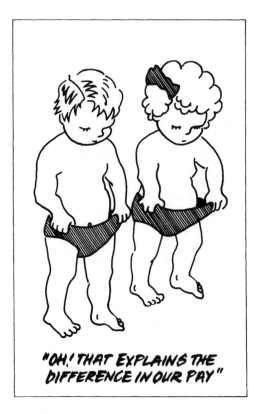

Figure 3.1 The differences between sex and gender

explain the differences in women's and men's lives. And yet, although gender might be distinguished from sex, they are also closely bound up together as gender roles are constructed differently for women and for men, and as gendered ideas about appropriate and desirable behaviour and aspirations come to be inscribed on sexed bodies. These social constructions are often very powerful and have material effects that shape the everyday lives of women and men. So, for example, ideas about the home, domesticity and part-time paid employment are often infused by ideas about femininity, identifying women rather than men as care-givers, domestic workers and consumers (Leslie, 1993). In contrast, ideas about full-time employment and citizenship beyond the home are often shaped by ideas about masculinity, reflected by notions of a male 'breadwinner' and the suitability of men rather than women to certain spheres of work and participation in public life (Massey, 1996).

Distinguishing between 'sex as biological' and 'gender as social' means that gender roles and relations are constructed and dynamic rather than fixed and static from birth. To think of gender as a social construction has been politically enabling, allowing gender roles and relations to be destabilised and resisted by both women and men. Anne Phillips describes the dynamic nature of gender:

> We are living through a time of major transformation in sexual relations: transformations that can be measured in the global feminization of the workforce, the rapid equalization between the sexes (at least in the richer countries) in educational participation and qualifications, and a marked increase in women's self-confidence and self-esteem that is probably the most lasting legacy of the contemporary women's movement. The changes cannot be attributed to feminism alone, and are often ambiguous in their effects; but even if the reshaping of gender relations is partial and deeply problematic, it would be hard not to notice this as a period of significant change. (Phillips, 1998: 1; also see Rowbotham, 1997, for a wide-ranging account of women's lives in Britain and the United States over the course of the twentieth century)

So, for example, it is now unusual for a man to be the sole 'breadwinner' in a family, and more women are in paid employment. And yet, although some gender roles and gender relations may have changed, others remain the same. While more women than ever before work outside the home, they remain disproportionately responsible for childcare and other domestic work and are more likely to be employed in a part-time capacity and on lower wages than men (Pratt and Hanson, 1995). Although more women are entering higher education, this occurs unevenly between different disciplines, with the natural sciences continuing to be largely dominated by men. Within a discipline such as geography, the gender balance among undergraduates is more likely to approach equality than the gender balance among postgraduates and academic staff (McDowell, 1990; Rose, 1993).

Many areas of feminist theory and politics continue to identify gender in terms of its difference from sex. But in recent years, the distinction between sex and gender has come to be questioned in important and challenging ways.

Inspired by the work of feminist theorists such as Judith Butler, Donna Haraway and Elspeth Probyn, it has been argued that it is impossible to think of biological sex without invoking ideas of gender and that to think in terms of gender necessarily invokes ideas of sexual difference (see, for example, Butler, 1990, 1993; Haraway, 1991; Probyn, 1993). In other words, ideas about a sexed identity as male or female are themselves gendered, while ideas about gender identity in terms of masculinity or femininity are inevitably rooted in ideas of biological sex. In different ways, Haraway, Butler and Probyn stress the *materiality* of sexed and gendered identities on a bodily scale (see, for example, Young, 1989). Rather than view biological sex as unproblematic, innate and taken for granted, they stress that ideas of biological sex are themselves constructed in gendered terms. At the same time, rather than view gender as a free-floating social construction, they argue that it is embodied, sexed and performed in material ways. Both Haraway and Butler seek to disrupt the *heterosexism* of distinguishing between sex and gender. To think of gender in terms of male and female roles is to remain within a 'heterosexist matrix', which normalises opposite-sex desire between men and women and marginalises other sexualities as different and as deviant (see Chapter 4 for further discussion).

Gender identities are not fixed, static and singular. Rather, gender identities vary over time and space and for different people in different contexts. Gender identities may be contested, resisted, transgressed and subverted. In many ways, gender identities are inherently geographical, positioning people in relation to one another and in relation to social, economic, cultural and political spheres of life in different ways. At the same time, different spaces and places are gendered in different ways. Within a company, for example, the boardroom and the typing pool might be gendered in quite distinct ways, and the centre of a city might be gendered differently during the day and late at night. According to McDowell and Sharp, more complex understandings of gender have been increasingly articulated in spatial terms. As they write,

> Gender is only one aspect of our identities as women and men, its significance varies across space and time, as does its very constitution and meaning, but we do not see this as reducing the materiality of these divisions in the daily lives of women across the globe. Increasingly, however, feminist scholars have come to recognise that there are multiple ways of 'doing gender', of being male and female, masculine and feminine that are more or less appropriate, more or less socially sanctioned in particular spaces and at different times. Interestingly, this complexity and the idea of the positionality and contextual nature of gendered identities has, in our view, placed geographical research, a spatial imagination, right at the centre of current feminist scholarship. (McDowell and Sharp, 1997: 2)

Before discussing the different ways in which geographers have analysed the relationships between gender, space and power, we will introduce the 'first-wave' and 'second-wave' of feminist politics in Britain and the United States, which laid the foundations for feminist geographical work from the late 1970s.

Feminist politics

Although the term 'feminist' was not widely used until the twentieth century, women have resisted male domination for as long as it has existed. But the political *organisation* of women – what is often referred to as the *women's movement* – dates from the nineteenth century. The first example of a women's movement developed from a meeting at Seneca Falls in New York State in 1848. The meeting was called to consider the 'social, civil, and religious condition and rights of woman' (quoted in Schneir, 1996: 76), and was organised by Elizabeth Cady Stanton and Lucretia Mott. The 300 people attending the meeting voted to adopt a 'Declaration of Sentiments and Resolutions' (see Box 3.1). It was significant that this meeting took place when it did, because, as Schneir writes, 'in 1848, in England, France, Germany, Austria and elsewhere, people were taking to the streets, seeking the fulfillment of liberal democratic rights proclaimed in the great documents of the French and American Revolutions and, in many instances, demanding new economic rights for workers. Presaging the political and social storms of the future, that very same year Marx and Engels penned and issued the *Communist Manifesto*' (Schneir, 1996: 77; see Chapter 2 for further discussion). At a time of revolutionary challenges to the accepted social order, the Seneca Falls Declaration represented the first organised campaign for women's rights.

Box 3.1 Declaration of Sentiments and Resolutions, Seneca Falls, 1848

The Declaration agreed at Seneca Falls was written by Elizabeth Cady Stanton and listed the following injustices suffered by women at the hands of men:

> He has never permitted her to exercise her inalienable right to the elective franchize ...
> He has made her, if married, civilly dead ...
> He has taken from her all right in property, even to the wage she earns ... becoming to all intents and purposes, her master ...
> He has so framed the laws of divorce ... as to be wholly regardless of the happiness of women ...
> He has monopolized nearly all the profitable employments ...
> He has denied her the facilities for obtaining a thorough education ...
> He has created a false public sentiment by giving to the world a different code of morals for men and women ... (quoted in Schneir, 1996: 78–79)

The 12 Resolutions included the following:

> That woman is man's equal ...
> That the same amount of virtue, delicacy, and refinement of behavior that is required of woman in the social state, should also be required of man, and the same transgressions should be visited with equal severity on both man and woman ...
> That it is the duty of the women of this country to secure to themselves their sacred right to the elective franchise ... (quoted in Schneir, 1996: 81–82)

Although an organised women's movement emerged earlier in the United States than in Britain, both developed in similar contexts, where women were barred from most institutions of higher education, did not have the vote, and lost their right to own property after marriage. According to Bolt, early British and American feminist movements were 'rooted in basically similar and encouraging social conditions. These included a shared heritage of Enlightenment ideas, expanding political rights and political toleration, an economy shifting to industrialisation and urbanisation, an influential middle class and a predominantly Protestant culture' (Bolt, 1993: 3). While the campaign for the vote was a vital part of 'first-wave' feminism, it should be seen in the context of a broader concern for women's rights in, for example, education, work, marriage, sexual freedom and financial self-sufficiency. Broadly speaking, 'first-wave' feminism was concerned with the rights of women to exist independently from their fathers, husbands or sons. In both countries, the women's movement was largely, but not exclusively, middle class and, although many feminists in both countries were closely associated with anti-slavery campaigns, the women's movement was dominated by white women (see Box 3.2, and Ferguson, 1992; Midgely, 1998; Ware, 1992). In Britain, 'first-wave' feminist politics were often closely tied to imperial politics, equating the inferior position of western women with the position of colonised women while at the same time reaffirming a belief in western superiority and 'civilisation' (see Chapter 5, and Burton, 1994; Bush, 1998; Ware, 1992).

Box 3.2 Sojourner Truth: *Ain't I a woman?*

Sojourner Truth (1795–1883) (see Figure 3.2.1) was born into slavery. When New York State emancipated its slaves in 1827, Sojourner Truth gained her freedom. After several years of domestic work, 'she felt that she had been called by the Lord to travel up and down the land testifying to the sins against her people. Dropping her slave name, Isabella, she took the symbolic name of Sojourner Truth. She spoke at camp meetings, private homes, wherever she could gather an audience. By midcentury she was well known in anti-slavery circles and a frequent speaker at abolitionist gatherings' (Schneir, 1996: 93). In 1850, she was the only black woman to attend the First National Woman's Rights Convention, held in Worcester, Massachusetts. The Convention passed a resolution describing the million and a half women still enslaved in the south as 'the most grossly wronged and foully outraged of all women' and declared that 'in every effort for an improvement in our civilization, we will bear in our heart of hearts the memory of the trampled womanhood of the plantation, and omit no effort to raise it to a share in the rights we claim for ourselves' (quoted in Schneir, 1996: 93). In 1851, Sojourner Truth attended a women's convention in Akron, Ohio, where she spoke about her position in society and her place as a black woman:

> That man over there says that women need to be helped into carriages, and lifted over ditches, and to have the best place everywhere. Nobody ever helps me into carriages, or over mud-puddles, or gives me any best place! And ain't I a woman? Look at me! Look at my arm! I have ploughed and planted, and gathered into barns, and no man could head me! And ain't I a woman? I could

Box 3.2 *continued*

Figure 3.2.1 Sojourner Truth

work as much and eat as much as a man – when I could get it – and bear the lash as well! And ain't I a woman? I have borne thirteen children, and seen them most all sold off to slavery, and when I cried out with my mother's grief, none but Jesus heard me! And ain't I a woman? (Sojourner Truth, quoted in Schneir, 1996: 94–95; also see hooks, 1982)

While the earliest and most fully developed women's movement was located in the United States from the mid to late nineteenth century, the first *militant* feminist campaign emerged in the United Kingdom in the early twentieth century and concentrated on women's suffrage, or the right to vote in parliamentary elections (see Box 3.3 for a chronology of women's rights in Britain). The American state of Wyoming was the first place to grant women the right to vote in 1869, and the first country to do so was New Zealand in 1893 (Miles, 1988). Some women gained the right to vote in Britain in 1918, but this was not extended to all women aged over 21 until 1928. In the United States, a constitutional amendment granted women the vote in 1920. Women in Australia, Denmark, Finland, Iceland, Norway and the USSR had all gained the right to vote before women in Britain and the United States (Miles, 1988).

In Britain, the militant suffragette campaign was most closely associated with the activities of the Women's Social and Political Union. The WSPU was founded in 1903 by Emmeline Pankhurst and her daughters Christabel, Sylvia and Adela, and differed from previous organisations in its direct action, in line with its slogan 'Deeds not words' (see Atkinson, 1996; Holton, 1996). The direct action of suffragettes included arson attacks, window smashing, chaining themselves to railings, and raiding the House of Commons. By 1911, direct action tactics had extended to include 'the firing and bombing of churches, private houses and public amenities. Golfing greens [were] attacked with acid, hundreds of letters burnt in post-boxes, buildings defaced with graffiti. All targets [were] chosen to avoid any loss of life' (Atkinson, 1996: 195). Alongside such direct action, there were marches and demonstrations and a well-coordinated publicity campaign making use of visual images and the newspaper *Votes for Women* (see Figures 3.2 and 3.3). Many suffragettes were imprisoned for their activities, and from 1909 many went on hunger strike to protest about their treatment in prisons such as Holloway Gaol. The state response was to force-feed suffragettes and, in 1913, to pass the Prisoners' Temporary Discharge for Ill-Health Act, which came to be known as the Cat and Mouse Act. Under this Act, suffragettes on hunger strike were released from prison only to be rearrested when they recovered. The Act failed in its aims because many women managed to evade rearrest and many committed more crimes when they were out of prison on special licence (Atkinson, 1996). The Suffragette Campaign was suspended when the First World War broke out in August 1914. In 1918, when the War ended, the Representation of the People Act was the first legislation in Britain that granted some women the right to vote.

After the vote was won for some women, 'first-wave' feminist politics largely lost its momentum in Britain and the United States. According to Schneir,

> the decline of feminism after the First World War is attributable at least in part to the eventual concentration of the women's movements in Britain and the United States on the single narrow issue of suffrage – which was won. Other factors that have been cited are the postwar economic depression, the growing influence of anti-feminist Freudianism, and the development in Germany and

Box 3.3 Women's rights in the United Kingdom

1792	Mary Wollstonecraft's book *A Vindication of the Rights of Woman* demanded women's equality in political, civil and economic life
1848	Female students admitted to the University of London
1857	*Matrimonial Causes Act* set up divorce courts. Women could sue for divorce but not on the grounds of adultery. Mothers' right of access to their children after divorce were extended
1864, 66, 69	*Contagious Diseases Acts* regulated prostitution in certain ports and garrison towns in an attempt to counter venereal diseases in the army and navy
1870	*Married Women's Property Act* allowed women to keep £200 of their earnings for the first time. Previously all of a wife's possessions belonged to her husband
1872	The London School of Medicine for Women opened
1873	*Custody of Infants Act* enabled all women, whether of a 'blemished character' or not, to gain access to their children after separation or divorce
1876	Medical schools opened to women
1882	*Married Women's Property Act* allowed women to own and administer their own property
1884	*Married Women's Property Act* stated that a woman was a separate and independent person, no longer a 'chattel' of her husband
1886	Repeal of the *Contagious Diseases Acts*
1894	*Local Government Act* allowed women to vote for local parish councils
1897	*The National Union of Women's Suffrage Societies* united groups that had existed since 1865 in a federation led by Millicent Garrett Fawcett
1903	*The Women's Social and Political Union* founded in Manchester by the Pankhurst family
1918	*Representation of the People Act* gave the vote to certain women over 30 and enabled women to become MPs. Countess Markiewicz was the first woman MP but, as an active member of Sinn Fein, she refused to take the oath of allegiance and was not allowed to take her seat in Parliament
1919	*Sex Disqualification Act* opened all the professions (except the church) to women. Nancy, Lady Astor became the first woman MP to take her seat in Parliament
1923	Women could be granted a divorce on the grounds of adultery alone
1928	*Representation of the People Act* gave the vote to all women over 21
1969	*Representation of the People Act* gave the vote to all men and women over 18
1970	*Equal Pay Act* legislated that men and women should be paid the same for the same work
1975	*Sex Discrimination Act* made it illegal to discriminate between men and women in employment, education, housing and services

Figure 3.2 Selling 'Votes for Women' in London

Source: Atkinson, 1988, 1996

the Soviet Union of authoritarian governments that fostered male supremacist values. ... What neither movement [in Britain nor the United States] managed to develop was an ideology, and without a clearly defined theoretical position it is difficult for any cause to weather the storms of disappointment and defeat or the sometimes equally devastating effects of victory. The suffrage struggle provided the women's movements with a program under which mass organization and even pride in sisterhood might flourish, but which could not supply the intellectual basis to sustain the moment. (Schneir, 1995: xix, xxv)

The period from 1930 to 1960 – between the first and second waves of feminist politics – has been described as the 'counter-revolution' (Millett, 1971). And yet, Schneir argues that the years after the end of the Second World War witnessed the first signs of a feminist reawakening. In 1945, the United Nations affirmed the equal rights of women and men and soon afterwards established a Commission on the Status of Women. Suffrage was granted or extended to

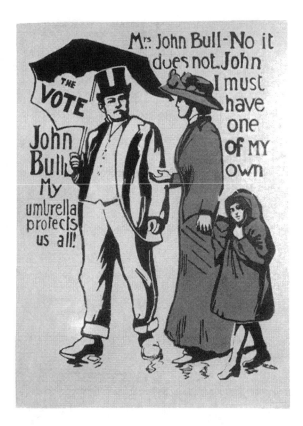

Figure 3.3 The women's suffrage campaign in Britain

women in countries that included France, Belgium, Italy, Venezuela, Japan and Korea. Simone de Beauvoir (1908–1986) published her influential book *The Second Sex* in 1949, in which she famously wrote that 'One is not born, but rather becomes, a woman' (quoted in Schneir, 1995: 3; see de Beauvoir, 1949). By the late 1960s, 'a vigorous feminist movement arose in the United States and rapidly spread to other countries. It was not really sudden, but it seemed so to those who had not been listening' (Schneir, 1995: xii).

In the United States, feminist politics followed two initially separate paths in the 1960s. First, from the early 1960s, the 'Women's Rights' movement included many professional women who campaigned for equality in employment, education, politics and the law through groups such as the National Organization for Women (see Box 3.4; Nicholson, 1997; Schneir, 1995). Second by the late 1960s, the 'Women's Liberation Movement' included many women involved in anti-Vietnam War campaigns and the Civil Rights movement who sought to liberate women from oppressive gender stereotypes and to overturn the sexism that existed throughout society. Different groups such as Redstockings, WITCH, The Feminists and New York Radical Feminists were formed to liberate women, and it was their members' 'vitality, daring, and creativity that gave feminism in our time its distinctive style and character' (Schneir, 1995: xii; see Box 3.4 and Figure 3.4).

Figure 3.4 'Women's lib' at the Statue of Liberty, New York, 1970

Box 3.4 The National Organization for Women and the Redstockings

Many different groups were formed in the United States in the 1960s to campaign for women's rights and women's liberation. Two important groups were the National Organization for Women and the Redstockings.

The National Organization for Women (NOW) was founded in 1966 by Betty Friedan in an attempt to ensure that the Equal Employment Opportunity Commission (EEOC) enforced its mandate to end sexual discrimination in paid employment. In its statement of purpose, NOW stated that:

> [T]he time has come for a new movement toward true equality for all women in America, and toward a fully equal partnership of the sexes, as part of the world-wide revolution of human rights now taking place within and beyond our national borders .... We believe that women will do most to create a new image of women by acting now, and by speaking out on behalf of their own equality, freedom, and human dignity ... in an active, self-respecting partnership with men. By so doing, women will develop confidence in their own ability to determine actively, in partnership with men, the conditions of their life, their choices, their future, and their society. (quoted in Schneir, 1995: 96, 102)

The Redstockings was a collective founded in New York City in 1969 by Shulamith Firestone and Ellen Willis. Its name plays on 'bluestockings', which was a derogatory term describing educated women in the late nineteenth and early twentieth centuries. The Redstockings Manifesto stated that:

> Women are an oppressed class. Our oppression is total, affecting every facet of our lives. We are exploited as sex objects, breeders, domestic servants, and cheap labor. We are considered inferior beings, whose only purpose is to enhance men's lives. Our humanity is denied. Our prescribed behavior is enforced by the threat of physical violence. Because we have lived so intimately with our oppressors, in isolation from each other, we have been kept from seeing our personal suffering as a political condition. (quoted in Schineir, 1995: 127)

The Redstockings coined famous new slogans and terms such as 'sisterhood is powerful', 'the personal is political', and 'consciousness raising'.

The women's liberation movement was closely associated with other radical politics that were changing American society in the 1960s. As Rowbotham writes,

> The use of nonviolence, the conviction that the means of struggle influenced the ends, the stress on self-emancipation as well as on rights, the claims on space and the demand to be part of the shaping of American society were all to be formative characteristics of a prolonged struggle to make a new politics. Women's liberation as a movement was to owe much to the Civil Rights and black movements. (Rowbotham, 1992: 258)

Women played an important part in the Civil Rights movement. As Schneir explains,

> Rosa Parks, a black woman from Montgomery, Alabama, set off a boycott of public transportation when she refused to give up her seat in a bus to a white passenger. Ella Baker, a seasoned NAACP [National Association for the Advancement of Colored People] leader who helped found the Student Nonviolent Coordinating Committee (SNCC), inspired many of the younger generation of civil rights activists. And Fannie Lou Hamer, a former Mississippi sharecropper, challenged the regular delegates from her state at the 1964 Democratic National Convention. Among the thousands of brave, idealistic young people, both black and white, who enlisted in the fight for racial justice, many were female. (Schneir, 1995: 89)

In 1964, the Civil Rights Act was passed to alleviate racial injustices. It was also the first national law in the United States that legislated for the equal employment opportunities of women and men (see Schneir, 1995: 71–73). As a result of this Act, the EEOC was established as a federal agency to hear complaints of workplace discrimination against women. But it was not until 1972 that this Commission could enforce its decisions in the courts:

> eventually female airline 'hostesses' won reversal of policies calling for mandatory dismissal at marriage or on reaching the age of thirty-five; the telephone company was forced to hire men as telephone operators and women as telephone-repair personnel; separate newspaper help-wanted listings under 'Male' or 'Female' headings were abolished; craft unions were opened to women who wished to work as carpenters, plumbers, and the like; women with the necessary skills were for the first time employed at the highest-paying industrial jobs; and charges of discrimination in hiring, pay, and promotion brought by corporate employees, university teachers, bank personnel, and other white-collar workers were favorably resolved. (Schneir, 1995: 73)

Although the rebirth of the modern feminist movement originated in the United States in the 1960s, it soon spread to many other countries (Miles, 1988). In Britain, the Hull Equal Rights Group was founded in 1968 to support a safety campaign after men were killed on a trawler. The National Joint Action Campaign for Women's Equal Rights was established by trade unionists after sewing machinists at Ford's Dagenham plant went on strike in 1968 for the right to be graded as skilled workers (Rowbotham, 1997; see Figure 3.5). By 1969, there were a range of campaigns against the portrayal of women as domestic and consumer sex objects, including a demonstration at the Ideal Home Exhibition and the disruption of Miss World contests in 1969 and 1970. The first National Women's Liberation Conference in Britain was held in 1970 and the first Women's Liberation Demonstration took place in London in 1971 (Rowbotham, 1997). From the 1960s onwards, in the United States, Britain and many other countries, 'second-wave' feminist politics sought to overturn the inferior position of women at work and at home and in law, politics

Figure 3.5 The struggle for equal pay: Ford Dagenham Machinists' Strike

and education. 'Consciousness-raising' led many women in different groups to share their experiences of life *as women*. Recognising that 'the personal is political', many feminists campaigned for women's control over their own bodies, often arguing for the legalisation of abortion, the greater availability of contraception and an equal level of sexual freedom with men (see Chapter 4 for more on the ways in which gay liberation developed alongside campaigns for women's liberation). Rather than view women merely in biological terms as potential mothers, such campaigns sought to detach sex from reproduction so that women as well as men could express their sexuality more fully and freely.

Since the 1960s, feminist politics have ranged across diverse issues that include violence against women, sexual harassment, sexual discrimination, paid employment, domestic work, motherhood, reproductive rights and media images of women. Unlike the 'first wave' of feminist politics, the 'second wave' has been characterised by the emergence of diverse feminist theories across many different disciplines that attempt to understand and to explain, as well as to change, gender inequalities. Over the course of the 1980s and 1990s, feminist politics have moved away from universalist and essentialist claims such as 'sisterhood is global' and a focus on 'woman' or even 'women' as an unproblematic category. In the next section, we will introduce geographies of gender from the late 1970s to the present by considering the inherent *spatiality* of both feminist politics and feminist theory.

Geographies of gender

Feminist politics are also *spatial* politics, resisting the confinement of women to certain spaces such as the home and campaigning for access to and equal participation within all spheres of life including work, politics and education. According to Carole Pateman, 'The dichotomy between the private and the public is central to almost two centuries of feminist writing and political struggle; it is, ultimately, what the feminist movement is about' (Pateman, 1989: 118). Feminist geographies are thus centrally important to feminist politics. Feminist geographies are diverse, spanning a wide range of theoretical, empirical, methodological and political concerns. They are also dynamic rather than static, changing over space and time and shaped by particular contexts and circumstances. This section will introduce feminist geographical work associated with liberal, socialist and poststructuralist perspectives. These perspectives often coexist, and the boundaries between them are much more fluid than fixed. Rather than view liberal, socialist and poststructuralist feminisms as clearly distinct and as neatly chronological, it is important to trace their connections and common ground while, at the same time, recognising their differences of emphasis and focus. As a central theme running through this section, we will focus on the ways in which these different perspectives have understood public and private space and gendered identities.

Liberal roots: adding women to geography

Liberalism is a well-established political tradition that seeks to promote individual autonomy, self-fulfilment and equality in a broader framework of human reason and rationality (see Tong, 1989, for an introduction to liberalism and liberal feminism). As Tong explains, the challenge set by liberalism is to create political, economic and social institutions that can maximise individual freedom, autonomy and self-fulfilment, without jeopardising the welfare of the community as a whole. Ideas about the differences between public and private spheres are central to liberal politics. In liberal terms, individuals have the right to an autonomous private sphere of home and family, which is separate from state intervention. Ideas about the nature of the public sphere differ. So, for example, *classical liberalism* views the ideal state as protecting civil liberties but not interfering in the economic market, while *welfare liberalism* views the ideal state as focusing on economic justice rather than civil liberties and intervening in positive ways through benefits and services such as the National Health Service, unemployment benefit and housing benefit in the UK. Liberal feminists have addressed the gender divisions between public and private space and argue for equality between men and women as autonomous and rational individuals.

The earliest liberal feminist writer is usually identified as Mary Wollstonecraft (1759–1797), who is best known for her book *A Vindication of the Rights of Woman*, which was first published in 1792 (Wollstonecraft, 1983). Wollstonecraft's book was published at a time of revolutionary uprisings in the United States and France and has been called 'the feminist declaration of independence' (Kramnick, 1983: 7): 'Wollstonecraft dared to take the liberal doctrine of inalienable human rights, a doctrine which was inflaming patriots on both sides of the Atlantic, and assume these rights for her own sex' (Kramnick, 1983: 7). In her book, she argued that women should benefit from the same education as men. Focusing on married, bourgeois women, she criticised their enforced confinement to the private sphere of the home, which served to deny them both autonomy and self-fulfilment. She described these women as always defined in relation to men – their fathers, husbands and sons – and therefore denied the individualism that was a central tenet of liberalism. Wollstonecraft argued that these women needed to be educated to enable them to move beyond the emotional, private sphere of the home into the rational, public sphere of citizenship. She represented this spatial division in vivid terms, writing that eighteenth-century married women were birds confined to cages, with nothing to do except plume themselves and 'stalk with mock majesty from perch to perch' (quoted in Tong, 1989: 14). Because of their spatial and social confinement in the private sphere, these women lacked health, liberty and 'virtue', which in liberal terms represented their inability to exercise reason and rationality (see Box 3.5 on the constraints experienced by women in urban space). Wollstonecraft's ideas offered a radical contrast to prevailing beliefs held about gender roles and education at the time. An influential account that Wollstonecraft critiqued was written by Jean-Jacques Rousseau, in which he contrasted an ideal male student, Emile, who studied the humanities, social and natural sciences, with an ideal female student, Sophie, whose education was restricted to the more feminine pursuits of music, art, fiction and poetry, all of which were thought to develop the qualities of a successful and desirable wife and mother. In Wollstonecraft's terms, Sophie and Emile would have had the same educational opportunities and would be equally valued in society.

In the nineteenth century, Wollstonecraft's ideas were extended by other liberal feminist writers who included Harriet Taylor and John Stuart Mill. Taylor and Mill wrote about the need not only for educational equality but also for the same civil liberties and economic opportunities for men and women, most famously in Mills's book *The Subjection of Women*, which was first published in 1869. The liberal feminist tradition has continued to be influential over the course of the twentieth century, and continues to strive for equality between women and men. In 1963, Betty Friedan published a book called *The Feminine Mystique*, which was to have a major impact on 'second-wave' feminist politics. For Friedan 'the feminine mystique' referred to the idea that a woman could only be satisfied, fulfilled and autonomous in the traditional roles of wife and mother. But she stressed that this was a 'mystique', because women in these roles were often dissatisfied and unfulfilled in their lives. Describing the life of an American suburban wife, Friedan wrote:

> As she made the beds, shopped for groceries, matched slipcover material, ate peanut butter sandwiches with her children, chauffeured Cub Scouts and Brownies, lay beside her husband at night – she was afraid to ask even of herself the silent question – 'Is this all?' (Friedan, quoted in Schneir, 1995: 50)

Box 3.5 Sexing the city

A wide range of feminist work has explored the ways in which urban space is gendered in distinctive ways. As Elizabeth Wilson writes,

> The relationship of women to cities has long preoccupied reformers and philanthropists. In recent years the preoccupation has been inverted: the Victorian determination to control working-class women has been replaced by a feminist concern for women's safety and comfort in city streets. But whether women are seen as a problem of cities, or cities as a problem for women, the relationship remains fraught with difficulty.
> (Wilson, 1992: 90)

In both historical and contemporary contexts, feminists have explored the ways in which urban space has been represented and experienced as dangerous for women (see, for example, Pain, 1991, 1997; Valentine, 1989; Walkowitz, 1992). Other work has focused on the constraints imposed on women's mobility and freedom within cities, with the figure of the *flâneur* as a key motif. The flâneur is often characterised as an archetypal hero of modernity, observing and consuming the flux and spectacle of modern urban life (see Pinder, in McDowell and Sharp, 1999, and Tester, 1994, for more on the flâneur). Feminists have explored the extent to which the flâneur is an exclusively male figure. Janet Wolff argues that the flâneur is inevitably masculine, and that there is no female equivalent *flâneuse* in the history of modernity (Wolff, 1990). She writes that modernity is represented in terms of the masculine occupation of public space – embodied by the flâneur – while the experiences of women in private space are largely overlooked. She cites the example of George Sand, a female French novelist in the nineteenth century who adopted a male name and masculine clothing to be able to occupy urban space. In contrast, Elizabeth Wilson challenges the idea that the flâneur is necessarily male and disrupts the separate spheres of public and private space (Wilson, 1991, 1992). She writes that women increasingly worked outside the home and that new places of leisure and consumption were designed to attract middle-class women. So, for example, Wilson sees shopping as a central part of *flânerie*: 'Although one could argue that shopping was for many women – perhaps the majority – a form of work rather than pleasure, at least for the leisured few it provided the pleasures of looking, socializing and simply strolling – in the department store, a woman, too, could become a flâneur' (Wilson, 1992: 101; also see Dowling, 1993, on department stores as feminised spaces of consumption).

For Friedan, such women had been increasingly confined to the private sphere of home and family so that jobs were available for men returning from the Second World War. At the same time, such women were viewed not only as home-makers but, crucially, as domestic consumers, fuelling the demand for new consumer goods (also see Dowling, 1993; Leslie, 1993). In Friedan's argument, women could only become autonomous and self-fulfilled individuals if they worked outside the home. But, as Tong explains,

> The Feminine Mystique failed to consider just how difficult it would be for even privileged women to combine marriage and motherhood with a career unless major structural changes were made within, as well as outside, the family. Like Wollstonecraft, Taylor and Mill before her, Friedan sent women out into the public realm without summoning men into the private domain. (Tong, 1989: 24)

In other words, even though women might move beyond the private sphere to enter paid employment, the gender division between public and private space remained largely intact.

Liberal feminist ideas have been extremely influential in arguing for women's rights, the need to value women as much as men, and the need to increase access for women to education, employment, the law and politics. But liberal feminist ideas have also been criticised on several fronts. Liberal feminists often refer to 'women' as an unproblematic category that remains undifferentiated by, for example, race, age, sexuality and class. Many liberal feminist ideas, including those of Wollstonecraft and Friedan, focus on white, bourgeois women without explicitly examining the difference that both whiteness and privilege make to the lives of such women. Women have always worked outside the home, and many are also employed as domestic workers in other people's homes (see, for example, Gregson and Lowe, 1994). As Schneir writes, The Feminine Mystique 'did not deal with the dissatisfactions of poor and often immigrant or minority women, who worked out of necessity rather than for personal fulfilment' (Schneir, 1995: 49). At the same time, liberal feminists have been criticised for seeking gender equality without structural changes in society. In other words, liberal feminists seek to 'add' women to institutions and paid employment that remain profoundly unequal. While individual women may benefit and may be able to achieve autonomy and self-fulfilment, the structures of unequal gender relations remain unchanged.

Liberal feminist ideas are inherently geographical, distinguishing between public and private spaces and between the position of women and men in society. Liberal feminist ideas have also been influential in geographical work, particularly in the late 1970s and early 1980s when the connections between geography and gender were first explored (see, for example, Ardener, 1981; Monk and Hanson, 1982; Tivers, 1978, 1985; WGSG, 1984; also see Bowlby et al., 1989). Clear parallels can be drawn between liberal feminist goals and the goals of early feminist geography, which included the right of women to be able to study and to teach as geographers and the importance of geographers studying the lives of women – of 'not excluding half the human in human geography' (Monk and Hanson, 1982). Liberal feminist ideas were highly influential for feminist geographers who sought to challenge the sexist bias of a subject that had largely overlooked the lives of women. As liberal feminists showed, geographical work had always been gendered – largely researched, written and taught by men and focusing on male, public spheres of life – but it had mostly ignored its gendered basis. Early feminist geographers played vital roles in revealing and resisting the masculinist basis both of geographical discourse and of geography as a discipline.

Resisting patriarchy and capitalism: socialist feminisms

While liberal feminists focus on the lives of individual women and the ways in which they could achieve autonomy and self-fulfilment, socialist feminists examine gender identity and subordination in more collective and structural terms, arguing that gender and class oppressions are intimately bound up together in capitalist societies. While marxist feminists view class as the main social relation to which gender is secondary, socialist feminists view class and gender as closely connected in all spheres of life (see Chapter 2 and Tong, 1989): 'Whereas socialist feminists believe that gender and class play an approximately equal role in any explanation of women's oppression, Marxist feminists believe that class ultimately better accounts for women's status and function(s)' (Tong, 1989: 39). Socialist feminists are concerned with the inter-dependence of class and gender inequalities under capitalism. As Tong writes, 'many contemporary socialist feminists have become convinced that living in a class society is not the only, or even primary, cause of women's oppression *as* women' (Tong, 1989: 174). Socialist feminist ideas have been centrally important in feminist geographical work, particularly over the course of the 1980s. Ideas of public and private space are important in socialist feminist work. But unlike liberal feminists, socialist feminists stress the critical connections between the private spaces of reproduction and the public spaces of production in a capitalist system and focus on women working within but also beyond the home.

Socialist feminists have theorised patriarchy and the connections between patriarchy and capitalism in different ways. As Linda McDowell explains,

> In its most general sense, the term patriarchy refers to the law of the father, the social control that men as fathers hold over their wives and daughters, but in its more specific usage within feminist scholarship patriarchy refers to the system in which men as a group are assumed to be superior to women as a group and so to have authority over them. (McDowell, in McDowell and Sharp, 1999: 196)

In geography, a debate about the nature of patriarchy appeared in *Antipode* in 1986. Jo Foord and Nicky Gregson argued that patriarchy is constituted by two universally necessary factors – biological reproduction and heterosexuality – and by two historically contingent factors – marriage and the nuclear family (Foord and Gregson, 1986). Their argument was forcefully critiqued, particularly because of its heterosexism and its abstraction (see Chapter 4 for more on heterosexism and heterosexist interpretations of gender). In a particularly influential and more materially grounded account, Sylvia Walby argues that men dominate and exploit women through the operation of six patriarchal structures: household production, paid employment, the state, male violence, sexuality, and cultural institutions (Walby, 1990). Each of these patriarchal structures shapes unequal social relations between men and women.

Socialist feminists have interpreted the connections between patriarchal and capitalist oppressions in different ways. So, for example, *dual systems theory* proposes that patriarchy and capitalism are distinct social relations that intersect in the oppression of women, while *unified systems theory* views patriarchy and capitalism as inseparable social relations that structure class and gender relations throughout society (McDowell, 1986; Tong, 1989). The work of socialist feminist geographers has been particularly important in explaining *where*, as well as why and how, gender and class oppressions take place. This work has tended to focus on the interdependence of geography, gender relations and economic development under capitalism. Rather than merely add women to geographical study, socialist feminist geographers focus on the structural roots of gender and class inequalities within a capitalist system. As Geraldine Pratt has written, the work of socialist feminist geographers has largely considered urban and regional scales of analysis, including work on the social and spatial separation of home and work (Pratt, 1994) (see, for example, MacKenzie and Rose, 1983; Pratt and Hanson, 1988, 1995), and gender divisions of labour within organisations (including Massey, 1994, 1996; Reimer, 1998). Socialist feminist geographers have also explored the spatial differentiation of gender relations and the uneven development of capitalism. On a regional scale, important work has analysed economic restructuring not only in spatial terms but also in terms of gender divisions. So, for example, geographers have traced the decline of employment in male-dominated manufacturing and the rise of a feminised service sector in advanced capitalist economies (McDowell, 1991). The feminisation of work refers not only to increased participation by women in paid employment, but also to the changing nature of employment and its increased flexibility. As Guy Standing explains,

> the types of work, labour relations, income, and insecurity associated with 'women's work' have been spreading, resulting not only in a notable rise in female labour force participation, but in a fall of men's employment, as well as a transformation – or feminization – of many jobs traditionally held by men. (Standing, 1989: 1077)

Many writers also explore the feminisation of work on a global scale, showing how and why the 'new international division of labour' is also a gendered division of labour (as discussed in Box 3.6).

Box 3.6 The international feminisation of work

Perhaps the most graphic example of the international feminisation of work exists in the employment of women in export-led manufacturing in less developed countries. Export-led manufacturing produces goods for export, often for multinational companies, and often in export processing zones and world market factories. Labour costs are kept as low as possible to ensure international competitiveness. Women make up the majority of workers in intensive, light manufacturing, producing goods such as clothing and textiles, toys and electronic goods. As Joekes explains, 'industrialization in the postwar

Box 3.6 *continued*

period has been as much *female* led as *export* led' (Joekes, 1987: 81; also see Elson and Pearson, 1981; Enloe, 1990). Women are disproportionately employed in export-led manufacturing because their labour is cheap. But, as Enloe shows, women's labour is not inherently cheaper than men's labour; it is rather *made* cheap:

> It has become commonplace to speak of 'cheap women's labour'. The phrase is used in public policy discussions as if cheapness were somehow inherent in women's work. In reality women's work is only as unrewarded or as low-paid as it is made to be. The international political economy works the way it does … in part because of the decisions which have cheapened the value of women's work. These decisions have first feminized certain home and workplace tasks – turning them into 'women's work' – and then rationalized the devaluation of that work. Without laws and cultural presumptions about sexuality, marriage and feminine respectability these transformations wouldn't have been possible. (Enloe, 1990: 160)

Enloe shows how three myths about the cheapness of women's work have material consequences for the lives of millions: first, that *women don't need to earn as much as men*, which results in lower wages for women; second, that *women have 'nimble fingers' or natural dexterity*, which means that their skilled work often goes unrewarded and unrecognised; and, third, that *women are docile*, which means that women are often prevented or discouraged from organising in trade unions or other political groups (Enloe, 1990).

Gender and difference: poststructuralist feminisms

Over the course of the 1990s, many feminist geographers have increasingly worked with poststructuralist ideas in their analyses of gendered spaces and the spatiality of gender. Poststructuralism represents a diverse field of work, which is concerned with the connections between power and the production of knowledge, the constitution and performativity of subjectivity, and the importance of difference (see Weedon, 1997, for a clear and helpful introduction). At its broadest, poststructuralism can offer 'a new way of analyzing constructions of meaning and relationships of power that called unitary, universal categories into question and historicized concepts otherwise treated as natural (such as man/ woman) or absolute (such as equality or justice)' (Scott, 1988a: 33). In other words, poststructuralism seeks to disrupt meanings that are assumed to be natural and taken for granted. Poststructuralist ideas also view power relations as diffuse throughout society, positioning individuals in relation to each other as well as in relation to the social body and working through, for example, language, discourse and a disciplinary gaze (as discussed more fully in Chapter 4).

The work of Michel Foucault has been particularly important in stimulating poststructuralist feminist work that considers gender in discursive terms, the links between gender and sexuality, and the ways in which power and knowledge are bound up together (see Chapter 4 for more on Foucault's work on the history of sexuality). The work of Jacques Derrida has also been very influential, proposing the *deconstruction* of binary oppositions – such as male/female and masculinity/femininity – that structure meaning and come to be assumed as natural (see Chapters 4 and 5 for further discussion of the need to overturn binary oppositions). For the purposes of this chapter, we will concentrate on poststructuralist ideas about discourse, difference and embodiment and their implications for understanding public and private space and gendered identities.

In general terms, 'discourse' refers to written and verbal communications but, informed by Foucault's work, it has come to acquire a more distinct meaning in poststructuralist thinking as 'the ensemble of social practices through which the world is made meaningful and intelligible to oneself and to others' (Gregory, 1994: 136). In Scott's terms, discourse represents 'a historically, socially, and institutionally specific structure of statements, terms, categories, and beliefs' (Scott, 1988a: 35). Particular knowledges of the world are produced and reproduced through representations and practices that often come to be naturalised, assumed and taken for granted. Feminist theorists have explored the ways in which gendered identities are discursively produced, through, for example, the media, law and education, and the ways in which masculine discourses come to be inscribed on male bodies and feminine discourses come to be inscribed on female bodies. At the same time, many poststructuralist critics resist the binary distinction between 'masculinity' and 'femininity', 'maleness' and 'femaleness', and suggest that identities or subjectivities are performed rather than naturally given (see Chapter 4 for more on ideas about performativity, sexuality and space). Although Foucault himself wrote little about gender, many feminists have extended his ideas to represent gender in discursive terms, to analyse the ways in which self-surveillance and self-discipline are gendered in distinctive ways, and to explore the production and regulation of sexed and gendered bodies (see, for example, Diamond and Quinby, 1988; Sawicki, 1991).

Poststructuralist feminist work has also challenged essentialist ideas about gender, which assert the category 'woman' or 'women' without tracing its complex differentiations. Rather than focus on 'women' as a singular and apparently unproblematic category, poststructuralist feminists seek to examine and to contextualise the complex and material interplay between, most commonly, gender, race, class, and sexuality. Unlike a liberal feminist focus on the essential differences and inequalities *between* men and women, poststructuralist feminists explore the differences *among* as well as between women and men (Barrett, 1989; Pratt and Hanson, 1994; Weedon, 1999). For example, Hazel Carby has written about slavery in the United States and the ways in which black women such as Sojourner Truth were dehumanised in contrast to white women. As she writes, white feminists in the first wave of American feminist politics were often *married* to white men, while black feminists were often *owned* by white men (Carby, 1982). Feminist politics and feminist theory has been increasingly challenged for its largely white, western, middle class and heterosexist underpinnings (see, for example, Box 3.7 on womanism; Collins,

1992; hooks, 1991; Mohanty, 1988; Probyn, 1993). So, for example, liberal feminist ideas about public and private spaces are increasingly recognised as more relevant in the lives of some women – particularly white, middle class, married women – than others. Feminist geographers have increasingly examined the *production* of public and private spaces, viewing them as relational, permeable and dynamic rather than static and fixed, and contextualising their importance for different women and men at different times and in different places (see, for example, Blunt and Rose, 1994).

Box 3.7 Womanism

Many black feminists and feminists of colour prefer the term 'womanism' to 'feminism' because the latter has been largely white, and largely uncritical of its whiteness. Alice Walker defines 'Womanist' in the following way:

> WOMANIST 1. From *womanish*. (Opp. of 'girlish,' i.e., frivolous, irresponsible, not serious.) A black feminist or feminist of color. From the black folk expression of mothers to female children, 'You acting womanish,' i.e., like a woman. Usually referring to outrageous, audacious, courageous or *willful* behaviour. Wanting to know more and in greater depth than is considered 'good' for one. Interested in grown-up doings. Acting grown up. Being grown up. Interchangeable with another black folk expression: 'You trying to be grown.' Responsible. In charge. *Serious*.

* * * * *

> 2. *Also*: A woman who loves other women, sexually and/or nonsexually. Appreciates and prefers women's culture, women's emotional flexibility (values tears as natural counterbalance of laughter), and women's strength. Sometimes loves individual men, sexually and/or nonsexually. Committed to survival and wholeness of entire people, male *and* female. Not a separatist, except periodically, for health. Traditionally universalist, as in: 'Mama, why are we brown, pink, and yellow, and our cousins are white, beige, and black?' Ans.: 'Well, you know the colored race is just like a flower garden, with every color flower represented.' Traditionally capable, as in: 'Mama, I'm walking to Canada and I'm taking you and a bunch of other slaves with me.' Reply: 'It wouldn't be the first time.'

* * * * *

> 3. Loves music. Loves dance. Loves the moon. *Loves* the spirit. Loves love and food and roundness. Loves struggle. *Loves* the Folk. Loves herself. *Regardless*.

* * * * *

> 4. Womanist is to feminist as purple to lavender.

Source: Walker, 1984: xi–xii

Many poststructuralist feminists represent gender identities in explicitly spatial terms, invoking spatial imagery that can include a rhetoric of mobility, positionality, marginality, and borderlands (Pratt, 1992; see Smith and Katz, 1993, who stress the importance of grounding such spatial images). Clear tensions exist between asserting and resisting gender identities, and these tensions are often articulated spatially in terms of situatedness, a politics of location, mobility and transgression (see, for example, McDowell, 1999; Pratt, 1984; Rich, 1986). In geography, poststructuralist feminist work has explored the spatiality of gender identities, as shown by studies of imperial women travellers. So, for example, the travels and writings of Mary Kingsley, who travelled in West Africa in the 1890s, have revealed the ways in which she was able to transgress the feminised domesticity of life in Victorian London because she was primarily identified as a white woman and able to share in imperial power while she was travelling away from home (Blunt, 1994; also see Frankenberg, 1993: 1, in which she analyses whiteness in spatial terms: 'First, whiteness is a location of structural advantage, of race privilege. Second, it is a "standpoint", a place from which white people look at ourselves, at others, and at society. Third, "whiteness" refers to a set of cultural practices that are usually unmarked and unnamed'; and geographical work on whiteness, which includes Bonnett, 1997, and Jackson, 1998). Poststructuralist feminists have also

Box 3.8 Bodily geographies

Many feminists are increasingly theorising identity in embodied ways, blurring the distinction often drawn between sex and gender and locating these and other identities in materially specific ways. As Robyn Longhurst suggests, the body represents an important new scale of analysis for feminist and other geographies (Longhurst, 1995, 1997; see McDowell, 1995; Nash, 1994; and Chapter 4 for examples of such work). Elizabeth Grosz details the complex ways in which bodies represent much more than biology:

> By 'body' I understand a concrete, material, animate organization of flesh, organs, nerves, and skeletal structure, which are given a unity, cohesiveness, and form through the psychical and social inscription of the body's surface. The body is, so to speak, organically, biologically 'incomplete'; it is indeterminate, amorphous, a series of uncoordinated potentialities that require social triggering, ordering, and long-term 'administration'. (Grosz, 1995: 104)

Bodily geographies are important on a number of different levels. First, feminist geographies resist the disembodied nature of much geographical work by challenging a mind–body split that comes to privilege the abstractions of the former over the lived experiences of the latter (Rose, 1993). Second, bodily geographies are produced, disciplined and administered by different power relations throughout society. As Adrienne Rich shows, bodies are a crucial site of gender politics, including: 'The politics of pregnability and motherhood. The politics of orgasm. The politics of rape and incest, of abortion, birth control, forcible sterilization. Of prostitution and marital sex. Of what had been named sexual liberation. Of prescriptive heterosexuality. Of lesbian existence' (Rich,

Box 3.8 *continued*

1986: 212–213). Third, tensions exist between asserting and resisting bodily geographies. Feminist politics have often focused on issues to do with bodies, such as male violence, abortion, contraception, pornography, maternity, eating disorders and self-defence. But women have all too often been defined and *confined* in terms of their bodies, as objects of desire and as actual or potential mothers. As Judith Butler writes,

> Somebody might well say: isn't it the case that certain bodies go to the gynaecologist for certain kinds of examination and certain bodies do not? And I would obviously affirm that. But the real question here is: to what extent does a body get defined by its capacity for pregnancy? Why is it pregnancy by which that body gets defined? ... Now it seems to me that, although women's bodies generally speaking are understood as capable of impregnation, the fact of the matter is that there are female infants and children who cannot be impregnated, there are older women who cannot be impregnated, there are women of all ages who cannot be impregnated, and even if they could ideally, that is not necessarily the salient feature of their bodies or even of their being women. ... It's a practical problem. If you are in your late twenties or your early thirties and you can't get pregnant for biological reasons, or maybe you don't want to, for social reasons ... *you are struggling with a norm that is regulating your sex*. (Butler, 1994: 33–34; emphasis added)

written about the ways in which identities are *embodied* in distinctive ways, suggesting a material and an intrinsically spatial approach to theorising gender and other identities (see Box 3.8 for an introduction to bodily geographies).

Gender and geographical knowledge

Liberal, socialist and poststructuralist traditions have all been influential in shaping feminist geographies of gender. Feminist geographical work is diverse, but shares common interests not only in the existence, production and representation of gendered spaces but also in the spatiality of gender and other identities. This final section will focus on the disciplinary challenges posed by feminist geography. It will begin by considering feminist attempts to rewrite histories of geography before examining the gendered production of geographical knowledge today.

Feminist work in both history and geography has largely moved from an early concern to 'add' women to historical and geographical studies to a more critical attempt to explain and to change the gendered production of historical and geographical knowledge (Scott, 1988b). An important area of work within feminist geography has been the attempt to rewrite histories of geography that not only include women but also explore the gendered production of geographical knowledge in more critical and challenging ways. Rather than narrate a singular 'geographical tradition', feminist geographical work has revealed different traditions in the history of geography and has revealed and resisted the erasure of women as both producers and subjects of geographical knowledge. While some feminist geographers have studied the roles of white, imperial women travellers in the production of geographical knowledge (see, for example, Blunt, 1994; Domosh, 1991; McEwan, 1998b), others have focused on women who were employed as geography teachers, researchers, technicians and journal editors (see, for example, Maddrell, 1997, on Marion Newbigin, who edited the *Scottish Geographical Magazine* from 1902 to 1934). As Box 3.9 shows, heated debates about admitting women as 'Fellows' of the Royal Geographical Society revolved around questions of scientific and social status.

Box 3.9 The Royal Geographical Society and Women Fellows

The Royal Geographical Society (RGS) was founded in London in 1830, with a membership of travellers, explorers, military and naval officers. In 1860, Lady Jane Franklin was awarded the Founder's Gold Medal for her 'devotion' in financing expeditions searching for her husband who died trying to locate the North West Passage. Mary Somerville was awarded the Victoria Medal in 1869 for her geographical work, but, as the RGS noted, 'in addition to her researches into the phenomena of the heavens and the earth, [she] has also excelled in the arts of painting, music and all feminine accomplishments' (quoted in Blunt, 1994: 150). Despite these two awards, women were not admitted as fellows of the RGS until 1893, and then not again until 1913.

In April 1893, 21 women – most of whom were travellers – had been elected as fellows. Even though the vast majority of the RGS were in favour of admitting women, a motion to carry on doing so was defeated at a special general meeting in July. Those women already elected could remain as fellows, but no more were admitted until 1913. In the 1890s, the debates about admitting women to the RGS revolved around ideas of scientific and social status. It was argued that the presence of women would undermine the scientific integrity of the RGS and that it was neither appropriate nor desirable for women to be involved in either travel or geography. As one fellow argued, 'if we had ladies here as a matter of right it would be more the object of the Council to provide for their amusement than for the progress of Scientific Research ... We have already magic lanterns and dissolving views, in a short time we should probably have a piano' (quoted in Blunt, 1994: 152). It was also claimed that admitting women to the RGS would undermine its social as well as scientific status, with one fellow arguing that 'I think the Society should not be a registry office for teachers and governesses and that kind of thing' (quoted in Blunt, 1994: 152). In May 1893, two letters to the *Times* represented both sides of the debate. An anonymous letter asked:

> Will it be alleged by any serious or sensible person that Mrs Bishop [a traveller and travel writer] is not a better Fellow, a better traveller, a better writer, and a more thorough Geographer than 19–20ths of the 3500 male Fellows? Can it be argued that an average woman is not as useful as a member of the Society as an average schoolmaster, or clergyman, or retired officer? (quoted in Blunt, 1994: 151)

But another letter claimed that women had nothing to contribute to geography:

> We contest *in toto* the general capability of women to contribute to scientific geographical knowledge. Their sex and training render them equally

Box 3.9 *continued*

> unfitted for exploration; and the genus of professional female globetrotters with which America has lately familiarized us is one of the horrors of the latter end of the nineteenth century. (quoted in Blunt, 1994: 152)
>
> The debates about admitting women fellows to the RGS in the 1890s reveal the links between imperial travel and geographical knowledge and the ways in which both were gendered in distinctive ways (see Chapter 5 for more on imperial travel and geography).

In her 1993 book, *Feminism and Geography*, Gillian Rose examines the gendered production of geographical knowledge, arguing that geography as a discipline is *masculinist* (Rose, 1993; and see the review essay by Hyndman, 1995). Although she writes that 'the master subject of geography is not only masculine but white, bourgeois and heterosexual' (Rose, 1993: 10), Rose concentrates on gender largely in isolation from race, class and sexuality. Rose adopts the term 'masculinist' from Michèle Le Doeuff, who describes as masculinist 'work which, while claiming to be exhaustive, forgets about women's existence and concerns itself only with the position of men' (quoted in Rose, 1993: 4). As Rose writes,

> Masculinist work claims to be exhaustive and it therefore assumes that no-one else can add to its knowledge; it is therefore reluctant to listen to anyone else. Masculinist work, then, excludes women because it alienates us in its choice of research themes, because it feels that women should not really be interested in producing geography, and also because it assumes that it is itself comprehensive. ... Masculinism can be seen at work not only in the choice of topics made by geographers, not only in their conceptual apparatus, not only in their epistemological claim to exhaustive knowledge, but also in seminars, in conferences, in common rooms, in job interviews. (Rose, 1993: 4)

Drawing on poststructuralist and psychoanalytic theories, Rose identifies the limits set by a geography that remains masculinist in theory and in practice; critiques the location of feminist geography within these limits; and suggests that the repositioning of feminist geography in paradoxical space (where there is a simultaneous and fluid occupation of both centre and margin, inside and outside) is a way to challenge these limits. In her critique of geographical knowledge as masculinist, Rose distinguishes between knowledge based in the social sciences and in the humanities. She contrasts a *social–scientific* masculinity, which 'claims complete access to a transparent and knowable world' (Rose, 1993: 61) with an *aesthetic* masculinity, which 'claims complete sensitivity to a mysterious yet crucial world' (Rose, 1993: 61). So, for example, Rose shows how humanistic geographers have revered the home as a site of intense experiences, memories and nostalgia, but have largely overlooked the ways in which homes are gendered in particular ways. Rose also traces the uneasy coexistence of social–scientific and aesthetic masculinities in cultural geography, whereby 'pleasure in the landscape is often seen as a threat to the scientific gaze, and it is argued that the geographer should not allow himself to be seduced by what he sees' (Rose, 1993: 72). Rose writes that 'cultural geography's erotics of knowledge' (Rose, 1993: 109) are masculinist and heterosexist, gazing on a feminised landscape in voyeuristic, distanced and disembodied ways (see Box 3.10 for more on feminist geographies of the gaze).

Box 3.10 Feminist geographies of the gaze

Ideas about vision, visual representation and 'the gaze' are inherently spatial, usually implying a distance between the observer and what is being observed and often involving different strategies of bounding or enframing what is seen (see, for example, Cosgrove, 1985, on the development of the 'landscape idea' as a way of seeing from the Renaissance onwards). Many feminists have analysed the ways in which fields of vision are gendered in distinctive ways. Such work is usually concerned with embodying and situating the gaze, in terms of both the observer and what or who is being observed. At the same time, feminists have studied the politics of representation, exploring how, why and where certain images are produced and their material effects on the lives of women (see, for example, Bonner *et al.*, 1992). Feminist work in film theory and art history has been particularly influential in embodying and contextualising the gendered nature of spectatorship and the gaze (see the essays by Doane and Mulvey in Erens, 1990; see Nochlin, 1991; Pollock, 1988, 1996; and see the feminist geographical work of Nash, 1994, 1996). For example, Griselda Pollock applies an explicitly spatial analysis to the paintings of two female Impressionists: Berthe Morisot (1841–1896) and Mary Cassatt (1844–1926). Both artists were actively involved in the artistic group now labelled Impressionist in Paris in the late nineteenth century. But their work is now less well known than that of their male contemporaries, such as Degas and Monet. Pollock traces the spaces of their art in three connected ways: first, she focuses on the spaces represented, which were very often interior, domestic spaces; second, she explores the spatial order within their paintings, showing how two spaces are often separated by, for example, a balcony or a balustrade, marking spaces that are open to women and men in different ways and positioning both the painter and the viewer in immediate proximity to the subject of the painting; and, third, she examines the social spaces of representation, studying where and what a woman was able to paint. As Box 3.5 suggests, clear limits were imposed on the mobility of bourgeois women in cities such as Paris in the nineteenth century. As another female artist wrote of her life in Paris at the same time:

> What I long for is the freedom of going about alone, of coming and going, ... of stopping and looking at the artistic shops, of entering churches and museums, of walking about old streets at night; that's what I long for; and that's the freedom without which one cannot become a real artist. Do you imagine that I get much good from what I see, chaperoned as I am, and when, in order to go to the Louvre, I must wait for my carriage, my lady companion, my family? (from the diary of Marie Baskirtseff, quoted by Pollock, 1988: 70)

Although Rose argues that geography is masculinist, feminist geographical work has had a considerable impact on the production of geographical

knowledge. Feminist geographers have contributed a great deal to methodological debates, stressing the importance of situating and embodying the
production of knowledge and attempting to negotiate the power relations
between researcher and the subject of research in sensitive and reflexive ways
(see, for example, the papers by England and Katz in the special edition of
Professional Geographer, 1994, on 'Women in the Field'). While Rose writes
that 'fieldwork is an example of geographical masculinities in action. Fieldwork is a performance which enacts some of the discipline's underlying masculinist assumptions about its knowledge of the world' (Rose, 1993: 65), other
feminist geographers have engaged in fieldwork 'precisely to critique, decon-
struct, and reconstruct a more responsible, if partial, account of what is
happening in the world' (Hyndman, 1995: 200; also see Sparke, 1996).
Feminist geographers have questioned the location of 'the field', and stressed
the importance of positionality. As Cindi Katz puts it, 'we are always already
in the field – multiple positioned actors, aware of the partiality of all our stories
and the ... boundaries drawn in order to tell them' (Katz, 1994: 67).

An increasing amount of work is also exploring masculinities, often
informed by feminist theories about gender power and identities (see, for
example, Connell, 1995; Dawson, 1994). Geographical work on masculinities
is diverse, spanning the masculinities of imperial geography (Driver, 1992; and
see Chapter 5), discourses of masculine adventure in children's fiction (Phillips,
1997), masculinity and fieldwork (Sparke, 1996), the cultural politics of masculinity (Jackson, 1991, 1994), masculinity and work (McDowell, 1995; Massey, 1996), and the links between national identities and masculinity (Nash,
1996). While feminist geographies have disrupted the masculinist truth-claims
of geographical knowledge, they have also inspired an increasing amount of
critical work that focuses on the constitution and articulation of masculinities
over space and time and in particular places and contexts. Although this chapter has focused on the dissident geographies of feminist work, considering inequalities between as well as among women and men, it is also important to
study the growing body of work on masculine spaces and identities and the
critical links between them.

Conclusions

Feminist politics are also spatial politics, seeking to overturn gender inequalities
between men and women in different places and spheres of life. This chapter
began by introducing different understandings of sex and gender and their links
and differences. Then we introduced feminist politics, focusing on 'first-wave'
feminist politics of the late nineteenth and early twentieth centuries and 'second-
wave' feminist politics since the 1960s and 1970s. In geographical terms, we
focused on early feminism in the United States, the emergence of a militant feminism in Britain in the early twentieth century, and the revitalisation of feminism
in the United States and in Britain alongside other liberation politics in the 1960s
and 1970s. Feminist politics have always been distinctively geographical, not

only varying over space, but also involving campaigns for the right of women to occupy particular places in, for example, paid employment, education, the law and politics. Since the late 1970s, and particularly over the course of the 1980s and 1990s, feminist geographers have sought to explain and to resist gender inequalities both in the world and in the discipline of geography. While the work of feminist geographers is diverse, important themes include the study of space as gendered and the spatiality of gender. The diversity of feminist geographical work was introduced in terms of liberal, socialist and poststructuralist approaches. In particular, we examined the ways in which these different approaches have analysed the gender divisions between public and private space and the constitution of gender identities. The final part of the chapter turned to the disciplinary challenges posed by feminist geography, considering feminist attempts to rewrite histories of geography and the gendered production of geographical knowledge today. Feminist geographies are dissident geographies, working to change the imbalances of gender power both in the world and in the discipline of geography. Feminist geographies will remain critically important for as long as such imbalances continue to shape differences not only between but also among women and men.

References

ARDENER, S. (ed.) (1981) *Women and Space: ground rules and social maps.* London: Croom Helm.

ATKINSON, D. (1988) *Suffragettes.* London: Museum of London.

ATKINSON, D. (1996) *The Suffragettes in Pictures.* Stroud: Sutton Publishing.

BARRETT, M. (1989) Some different meanings of the concept of 'difference': feminist theory and the concept of ideology. In E. Meese and A. Parker (eds), *The Difference Within: feminism and critical theory.* Amsterdam: John Benjamin.

BLUNT, A. (1994) *Travel, Gender and Imperialism: Mary Kingsley and West Africa.* New York: Guilford.

BLUNT, A. and ROSE, G. (eds) (1994) *Writing Women and Space: colonial and postcolonial geographies.* New York: Guilford.

BOLT, C. (1993) *The Women's Movements in the United States and Britain from the 1790s to the 1920s.* New York: Harvester Wheatsheaf.

BONNER, F. *et al.* (eds) (1992) *Imagining Women: cultural representations and gender.* Cambridge: Polity.

BONNETT, A. (1997) Geography, 'race' and Whiteness: invisible traditions and current challenges. *Area,* **29**: 193–199.

BOWLBY, S. *et al.* (1989) The geography of gender. In R. Peet and N. Thrift (eds), *New Models in Geography,* Vol. 2. London: Unwin Hyman.

BURTON, A. (1994) *Burdens of History: British feminists, Indian women, and imperial culture, 1865–1915.* Chapel Hill: University of North Carolina Press.

BUSH, B. (1998) 'Britain's conscience on Africa': white women, race and imperial politics in inter-war Britain. In C. Midgely (ed.), *Gender and Imperialism*. Manchester: Manchester University Press, 200–223.

BUTLER, J. (1990) *Gender Trouble: feminism and the subversion of identity*. New York: Routledge.

BUTLER, J. (1993) *Bodies that Matter: on the discursive limits of 'sex'*. New York: Routledge.

BUTLER, J. (1994) Gender as performance: an interview with Judith Butler. *Radical Philosophy*, **67**: 32–39.

CARBY, H. (1982) White women listen! Black feminism and the boundaries of sisterhood. In Centre for Contemporary Cultural Studies, *The Empire Strikes Back: race and racism in 70s Britain*. London: Hutchinson, 212–235.

COLLINS, P.H. (1992) *Black Feminist Thought: knowledge, consciousness and the politics of empowerment*. London: Routledge.

CONNELL, R.W. (1995) *Masculinities*. Berkeley: University of California Press.

COSGROVE, D. (1985) Prospect, perspective and the evolution of the landscape idea. *Transactions of the Institute of British Geographers*, **10**: 45–62.

DAWSON, G. (1994) *Soldier Heroes: British adventure, empire and the imaginings of masculinities*. London: Routledge.

DE BEAUVOIR, S. (1972) [1949] *The Second Sex*. Harmondsworth: Penguin.

DIAMOND, I. and QUINBY, L. (eds) (1988) *Feminism and Foucault*. Boston, MA: Northeastern University Press.

DOANE, M.-A. (1990) Film and the masquerade: theorizing the female spectator. In P. Erens (ed.), *Issues in Feminist Film Criticism*. Bloomington: Indiana University Press.

DOMOSH, M. (1991) Toward a feminist historiography of geography. *Transactions of the Institute of British Geographers*, **16**: 95–104.

DOWLING, R. (1993) Femininity, place and commodities: a retail case study. *Antipode*, **25**: 295–319.

DRIVER, F. (1992) Geography's empire: histories of geographical knowledge. *Environment and Planning D: Society and Space*, **10**: 23–40.

DUMAYNE-PEATY, L. and WELLENS, J. (1998) Gender and physical geography in the United Kingdom. *Area*, **30**: 197–206.

DUNCAN, N. (1996) *BodySpace: destabilizing geographies of gender and sexuality*. London: Routledge.

ELSON, D. and PEARSON, P. (1981) Nimble fingers make cheap workers: an analysis of women's employment in Third World export manufacturing. *Feminist Review*, **7**: 87–107.

ENGLAND, K. (1994) Getting personal: reflexivity, positionality and feminist research. *Professional Geographer*, **46**: 80–89.

ENLOE, C. (1990) *Bananas, Beaches and Bases: making feminist sense of international politics*. Berkeley: University of California Press.

ERENS, P. (1990) *Issues in Feminist Film Criticism*. Bloomington: Indiana University Press.

FERGUSON, M. (1992) *Subject to Others: British women writers and colonial slavery, 1670–1834.* London: Routledge.

FOORD, J. and GREGSON, N. (1986) Patriarchy: towards a reconceptualisation. *Antipode*, **18**: 186–211.

FRANKENBERG, R. (1993) *White Women: race matters.* London: Routledge.

FRIEDAN, B. (1984) [1963] *The Feminine Mystique.* New York: Dell.

GREGORY, D. (1994) Discourse. In R. Johnston, D. Gregory and D. Smith (eds) *The Dictionary of Human Geography* (3rd edn). Oxford: Blackwell, 136.

GREGSON, N. and LOWE, M. (1994) *Servicing the Middle Classes: class, gender and waged domestic labour.* London: Routledge.

GROSZ, E. (1995) *Space, Time and Perversion: essays on the politics of bodies.* London: Routledge.

HARAWAY, D. (1991) *Simians, Cyborgs and Women: the reinvention of nature.* London: Free Association Books.

HOLTON, S.S. (1996) *Suffrage Days: stories from the women's suffrage movement.* London: Routledge.

HOOKS, B. (1982) *Ain't I a Woman? Black Women and Feminism.* London: Pluto.

HOOKS, B. (1991) *Yearning: race, gender and cultural politics.* Boston, MA: South End Press.

HYNDMAN, J. (1995) Solo feminist geography: a lesson in space. *Antipode*, **27**: 197–207.

JACKSON, P. (1991) The cultural politics of masculinity: towards a social geography. *Transactions of the Institute of British Geographers*, **16**: 199–213.

JACKSON, P. (1994) Black male: advertising and the cultural politics of masculinity. *Gender, Place and Culture*, **1**: 49–59.

JACKSON, P. (1998) Constructions of 'whiteness' in the geographical imagination. *Area*, **30**: 99–106.

JOEKES, S. (1987) *Women in the World Economy.* Oxford: Oxford University Press.

JONES, J.P., NAST, H. and ROBERTS, S. (eds) (1997) *Thresholds in Feminist Geography: difference, methodology, representation.* Lanham: Rowman and Littlefield.

KATZ, C. (1994) Playing the field: questions of fieldwork in geography. *Professional Geographer*, **46**: 67–72.

KRAMNICK, M.B. (1983) Introduction. In M. Wollstonecraft, *Vindication of the Rights of Woman.* Harmondsworth: Penguin.

LESLIE, D. (1993) Femininity, post-Fordism and the 'new traditionalism'. *Environment and Planning D: Society and Space*, **11**: 689–708.

LONGHURST, R. (1995) The body and geography. *Gender, Place and Culture*, **2**: 97–105.

LONGHURST, R. (1997) (Dis)embodied geographies. *Progress in Human Geography*, **21**: 486–501.

LOVENDUSKI, J. and RANDALL, V. (1993) *Contemporary Feminist Politics: women and power in Britain.* Oxford: Oxford University Press.

MACKENZIE, S. and ROSE, D. (1983) Industrial changes, the domestic economy and home life. In J. Anderson, S. Duncan and R. Hudson (eds), *Redundant Spaces: industrial decline in cities and regions*. London: Macmillan.

MADDRELL, A. (1997) Scientific discourse and the geographical work of Marion Newbigin. *Scottish Geographical Magazine*, **113**, 33–41.

MASSEY, D. (1994) *Space, Place and Gender*. Cambridge: Polity.

MASSEY, D. (1996) Masculinity, dualisms and high technology. In N. Duncan (ed.), *BodySpace*. London: Routledge, 109–126.

MCDOWELL, L. (1986) Beyond patriarchy: a class-based explanation of women's subordination. *Antipode*, **18**: 311–321.

MCDOWELL, L. (1990) Sex and power in academia. *Area*, **22**: 323–332.

MCDOWELL, L. (1991) Life without father and Ford: the new gender order of post-Fordism. *Transactions of the Institute of British Geographers*, **16**: 400–419.

MCDOWELL, L. (1995) Body work: heterosexual gender performances in city workplaces. In D. Bell and G. Valentine (eds), *Mapping Desire*. London: Routledge, 75–98.

MCDOWELL, L. (1999) *Gender, Identity and Place: understanding feminist geographies*. Cambridge: Polity.

MCDOWELL, L. and SHARP, J. (1997) *Space, Gender and Knowledge: feminist readings*. London: Arnold.

MCDOWELL, L. and SHARP, J. (1999) *A Feminist Glossary of Human Geography*. London: Arnold.

MCEWAN, C. (1998a) Gender, science and physical geography in nineteenth-century Britain. *Area*, **30**: 215–224.

MCEWAN, C. (1998b) Cutting power lines within the palaces? Countering paternity and eurocentrism in the 'geographical tradition'. *Transactions of the Institute of British Geographers*, **23**: 371–384.

MIDGELY, C. (ed.) (1998) *Gender and Imperialism*. Manchester: Manchester University Press.

MILES, R. (1988) *The Women's History of the World*. London: Paladin.

MILL, J.S. (1984) [1869] On the subjection of women. In J.M. Robson (ed.), *Collected Works of John Stuart Mill*. London: Routledge and Kegan Paul, 259–340.

MILLETT, K. (1971) *Sexual Politics*. London: Rupert Hart-Davis.

MOHANTY, C.T. (1988) Under Western eyes: feminist scholarship and colonial discourses. *Feminist Review*, **30**: 61–102.

MONK, J. and HANSON, S. (1982) On not excluding half of the human in human geography. *Professional Geographer*, **34**: 11–23.

MULVEY, L. (1990) Visual pleasure and narrative cinema. In P. Erens (ed.), Issues in *Feminist Film Criticism*. Bloomington: Indiana University Press.

NASH, C. (1994) Remapping the body/land: new cartographies of identity, gender and landscape in Ireland. In A. Blunt and G. Rose (eds), *Writing Women and Space: colonial and postcolonial geographies*. New York: Guilford, 227–50.

NASH, C. (1996) Men again: Irish masculinity, nature, and nationhood in the early twentieth century. *Ecumene*, **3**: 427–452.

NICHOLSON, L. (ed.) (1997) *The Second Wave: a reader in feminist theory*. New York: Routledge.

NOCHLIN, L. (1991) *Women, Art and Power and Other Essays*. London: Thames and Hudson.

PAIN, R. (1991) Space, sexual violence and social control: integrating geographical and feminist analyses of women's fear of crime. *Progress in Human Geography*, **15**: 415–431.

PAIN, R. (1997) Social geographies of women's fear of crime. *Transactions of the Institute of British Geographers*, **22**: 231–44.

PATEMAN, C. (1989) *The Disorder of Women: democracy, feminism and political theory*. Stanford, CA: Stanford University Press.

PHILLIPS, A. (ed.) (1998) *Feminism and Politics*. Oxford: Oxford University Press.

PHILLIPS, R. (1997) *Mapping Men and Empire: a geography of adventure*. London: Routledge.

PINDER, D. (1999), Flâneur/flâneuse. In L. McDowell and J. Sharp (eds), *A Feminist Glossary of Human Geography*. London: Arnold, 95–96.

POLLOCK, G. (1988) *Vision and Difference: femininity, feminism and the histories of art*. London: Routledge.

POLLOCK, G. (ed.) (1996) *Generations and Geographies in the Visual Arts: feminism and the histories of art*. London: Routledge.

PRATT, G. (1992) Spatial metaphors and speaking positions. *Environment and Planning D: Society and Space*, **10**: 241–244.

PRATT, G. (1994) Feminist geographies. In R. Johnston, D. Gregory and D. Smith (eds), *The Dictionary of Human Geography* (3rd edn). Oxford: Blackwell, 192–196.

PRATT, G. and HANSON, S. (1988) Gender, class and space. *Environment and Planning D: Society and Space*, **6**: 15–35.

PRATT, G. and HANSON, S. (1994) Geography and the construction of difference. *Gender, Place and Culture*, **1**: 5–30.

PRATT, G. and HANSON, S. (1995) *Gender, Work and Space*. London: Routledge.

PRATT, M.B. (1984) Identity: skin blood heart. In E. Burkin, M.B. Pratt and B. Smith (eds), *Yours in Struggle: three feminist perspectives on anti-semitism and racism*. Ithaca: Firebrand Books.

PROBYN, E. (1993) *Sexing the Self: gendered positions in cultural studies*. London: Routledge.

REIMER, S. (1998) Working in a risk society. *Transactions of the Institute of British Geographers*, **23**: 116–127.

RICH, A. (1986) Towards a politics of location. In A. Rich (ed.), *Blood, Bread and Poetry: selected prose, 1979–1985*. London: Virago.

ROSE, G. (1993) *Feminism and Geography: the limits of geographical knowledge*. Cambridge: Polity.

ROWBOTHAM, S. (1992) *Women in Movement: feminism and social action.* New York: Routledge.

ROWBOTHAM, S. (1997) *A Century of Women: the history of women in Britain and the United States.* London: Viking.

SAWICKI, J. (1991) *Disciplining Foucault: feminism, power and the body.* New York: Routledge.

SCHNEIR, M. (ed.) (1995) *The Vintage Book of Feminism: the essential writings of the contemporary women's movement.* London: Vintage.

SCHNEIR, M. (ed.) (1996) *The Vintage Book of Historical Feminism.* London: Vintage.

SCOTT, J. (1988a) Deconstructing equality-versus-difference: or, the uses of poststructuralist theory for feminism. *Feminist Studies,* **14**: 33–50.

SCOTT, J. (1988b) *Gender and the Politics of History.* New York: Columbia.

SMITH, N. and KATZ, C. (1993) Grounding metaphor: towards a spatialized politics. In M. Keith and S. Pile (eds), *Place and the Politics of Identity.* London: Routledge.

SPARKE, M. (1996) Displacing the field in fieldwork: masculinity, metaphor and space. In N. Duncan (ed.), *BodySpace: destabilizing geographies of gender and sexuality.* London: Routledge.

STANDING, G. (1989) Global feminization through flexible labour. *World Development,* **17**: 1077–1095.

TESTER, K. (ed.) (1994) *The Flâneur.* London: Routledge.

TIVERS, J. (1978) How the other half lives: the geographical study of women. *Area,* **10**: 302–306.

TIVERS, J. (1985) *Women Attached: the daily activity patterns of women with young children.* London: Croom Helm.

TONG, R. (1989) *Feminist Thought: a comprehensive introduction.* Boulder, CO: Westview Press.

VALENTINE, G. (1989) The geography of women's fear. *Area,* **21**: 385–390.

WALBY, S. (1990) *Theorising Patriarchy.* Oxford: Blackwell.

WALKER, A. (1984) *In Search of Our Mothers' Gardens.* London: The Women's Press.

WALKOWITZ, J. (1992) *City of Dreadful Delight: narratives of sexual danger in late Victorian London.* London: Virago.

WARE, V. (1992) *Beyond the Pale: white women, racism and history.* London: Verso.

WEEDON, C. (1997) *Feminist Practice and Poststructuralist Theory.* Oxford: Blackwell.

WEEDON, C. (1999) *Feminism, Theory and the Politics of Difference.* Oxford: Blackwell.

WILSON, E. (1991) *The Sphinx in the City: urban life, the control of disorder, and women.* London: Virago.

WILSON, E. (1992) The invisible flâneur. *New Left Review,* **191**: 90–110.

WOLFF, J. (1990) *Feminine Sentences: essays on women and culture.* Cambridge: Polity.

WOLLSTONECRAFT, M. (1983) [1792] *A Vindication of the Rights of Woman.* Harmondsworth: Penguin.

WOMEN AND GEOGRAPHY STUDY GROUP (1984) *Geography and Gender: an introduction to feminist geography.* London: Hutchinson.

WOMEN AND GEOGRAPHY STUDY GROUP (1997) *Feminist Geographies: explorations in diversity and difference.* Harlow: Longman.

YOUNG, I.M. (1989) Throwing like a girl. In J. Allen and I.M. Young (eds), *The Thinking Muse: feminism and modern French philosophy.* Bloomington: University of Indiana Press.

4

Sexual Orientations: Geographies of Desire

> Writing about sex has been one of the critical ways in which sexuality has emerged as a field of exploration, a terrain of battle, a continent of knowledge, a configuration of fear and desire – as well as a sphere of resistance and identity. (Weeks, 1991: 3)

Sexuality

Sexuality is, and always has been, an important part of human life (see Figure 4.1). Sexuality is an important way in which people imagine, identify and represent themselves and it affects how they are imagined, identified and represented by others. Sexuality also influences how people relate to others, not only as lovers, but also as friends, family and colleagues. But sexuality is not just about relationships between individuals. Rather, people are positioned in different ways in society because of their sexuality. Sexuality is a social relation that includes but also extends far beyond individual desires. As a social relation, sexuality is diverse, dynamic and contested, varying over time and space as different desires are not only articulated and fulfilled but also regulated, repressed and resisted. In many places, the dominance of heterosexuality represents the opposite-sex attractions of men and women as the social norm, against which the same-sex, homosexual, desire of lesbians and gay men and bisexual desire for both men and women are often seen to be abnormal, unnatural or deviant. As a result, many people are subject to homophobic violence, discrimination and marginalisation. Dissident sexual politics have played a crucial role in resisting homophobic oppression and in striving for the freedom to express and to fulfil different desires in the face of heterosexism. Although sexuality is and always has been central to human life, it is only in recent years that geographers have begun to study it, tracing the critical connections between sexuality and space and challenging the heterosexism that exists both within and beyond the academy.

Figure 4.1 Medieval lovers

This chapter begins by examining the dominance of heterosexuality – 'compulsory heterosexuality' – in greater detail. Then it will turn to the links between sex, power and knowledge, tracing a 'science of sex' that emerged from the nineteenth century, which defined sexuality as a legitimate object of study. Psychoanalysis has been particularly important in describing and attempting to explain human sexualities. The work of Sigmund Freud and Jacques Lacan will be introduced because it has provided many points of departure for critical work about sexuality. Michel Foucault has inspired a great deal of work on sexuality as discourse, and the geographical as well as historical specificity of sexual discourses will be outlined. After contextualising the links between sexuality, power and knowledge, the chapter will turn more explicitly to focus on dissident sexual politics in the twentieth century, concentrating on gay liberation campaigns in Britain and the United States. While the work of many geographers was inspired by and contributed to class and gender politics since the 1970s, it took longer for sexuality to appear on the geographical map, first appearing in the early 1980s but attracting more attention only over the course of the 1990s. The extent to which geographers were and are 'squeamish' (McNee, 1984) about studying sexuality will be explored. Sexual geographies pose important challenges not only to understand and to resist 'compulsory heterosexuality' in society but also in the discipline of geography. The work of geographers such as David Bell, Jon Binnie, Lawrence Knopp and Gill Valentine addresses not only the discursive connections between space and sexuality in the world, but also their disciplinary implications for geography and geographers. The chapter will end by exploring sexualities over space and the links between sexual spaces and sexual identities.

Challenging 'compulsory heterosexuality'

The term 'compulsory heterosexuality' has been most closely associated with the work of Adrienne Rich, who first used it in 1980 to describe the dominance of heterosexual relations and the ways in which heterosexuality has come to be naturalised in society (Rich, 1984; also see the work of Judith Butler, for whom a 'heterosexual matrix' designates 'that grid of cultural intelligibility through which bodies, genders, and desires are naturalized': Butler, 1990: 151). As Gill Valentine explains,

> The dominant form of sexuality in modern Western culture is heterosexuality, despite the fact that same-sex relationships have occurred throughout time and across different societies and cultures with varying degrees of acceptability and frequency (D'Emilio and Freedman, 1988). Such dominance is attributed to the fact that men and women have a biological instinct to reproduce. 'Normal sex' is therefore defined as potentially reproductive involving penetration with a penis and is usually assumed to take place within a monogamous relationship. (Valentine, 1993a: 395)

Through different channels and institutions such as the media, the law, education and religion, heterosexuality is so pervasive in society that it comes to be assumed as natural. Moreover, the dominance of heterosexuality comes to structure relations between men and women throughout society, fixing gender difference between the sexes. By linking sex and sexuality to reproduction and by confining it to the private space of the home and the unit of the nuclear family, heterosexuality shapes gender as well as sexual relations between men and women. So, for example, women are often defined in terms of their ability to be mothers, assuming their reproductive role in heterosexual society, often at the expense of their own desires (Butler, 1994; Russett, 1989). As discussed in Chapter 3, marxists and socialist feminists have stressed the importance of the reproductive sphere to capitalist economies, showing the centrality of heterosexuality to economic and political as well as social and cultural life. Many poststructuralist feminists have challenged the distinction drawn between sex as biological and gender as socially constructed, arguing that such a distinction remains heterosexist. Feminists such as Judith Butler and Elspeth Probyn instead show how understandings of sex and gender are bound up together, as discussed in Chapter 3. Heterosexuality is such a dominant social relation and so taken for granted that, in many ways, it appears to be paradoxically asexual. For example, the heterosexual nuclear family is widely represented as the norm to aspire to and a model for communities and nations to support and, indeed, to emulate (Parker et al., 1992). And yet, the sexual underpinnings of the heterosexual nuclear family – what goes on in the parental bedroom – are rarely mentioned.

But sexuality represents more than private sexual acts and its importance in human life is not confined to private homes and bedrooms. Sexuality extends far beyond domestic limits to encompass many other spaces of human interaction. As Valentine continues,

> heterosexuality is clearly the dominant sexuality in most everyday environments, not just private spaces, with all interactions taking place between sexed actors. However, such is the strength of the assumption of the 'naturalness' of heterosexual hegemony, that most people are oblivious to the way it operates as a process of power relations in *all* spaces. (Valentine, 1993a: 396)

In other words, heterosexuality is so dominant that it is taken for granted and assumed to be natural throughout society (see Figure 4.2). Moreover, heterosexuality is so naturalised that often neither its sexual basis nor its pervasive power are remarked upon. As Heidi Nast explains,

> heterosexuality is constructed as benign and/or asexual. By benign, I mean that heterosex's normative public expressions are seen as innocent, natural, or unremarkable: (typically white) hetero-couples kissing in parks; placing a public advert in a local newspaper for a hetero-mate; the public predominance of heterosexual dating agencies; and promotional tourism images of affectionate heterosexual couples, often scantily clad, sipping pina coladas

Figure 4.2 Challenging compulsory heterosexuality

or wading through blue waters with small children in tow, are simply not perceived as racy or even sexy …. In contrast, homosexual couples carrying out, or portrayed doing the same kinds of activities in public arenas (a ClubMed tourism advertisement showing a racially 'mixed' lesbian couple, scantily clad and romantically intertwined) are seen as <u>sexed and deviant</u>. (Nast, 1998: 192)

By being surrounded by heterosexual images in all spheres of life, male–female desires come to be naturalised and taken for granted to such an extent that the sexual basis of heterosexuality is effectively erased. In contrast, images of same-sex desire are invested with sexual meanings that are seen to differ from the social norm. The questions in Box 4.1 parody the ways in which heterosexual relations are assumed as normal rather than interrogated as different.

Box 4.1 Disrupting 'compulsory heterosexuality'

Heterosexuality has become naturalised and taken for granted as a social norm against which same-sex desires are often seen as different and deviant. As a result, many lesbians and gay men are put into situations where they have to defend their sexuality and where they have to explain and justify their difference from the social norm. The following questions parody homophobic questions that ask people to defend their sexuality. Instead of being addressed to lesbians or gay men, the questions are addressed to heterosexuals, and are based on 'heterophobic' rather than homophobic attitudes (adapted from the work of Dr Alan Malyon).

1. What do you think caused your heterosexuality?

2. When and how did you first decide that you were heterosexual?

3. Is it possible that your heterosexuality is just a phase you'll grow out of?

4. If you've never slept with a person of the same sex how do you know you wouldn't prefer that?

5. Your heterosexuality doesn't offend me as long as you leave me alone, but why do so many heterosexuals try to seduce others into that orientation?

6. Why must heterosexuals be so blatant, making a public spectacle of their heterosexuality? Can't you just be what you are and keep it quiet?

'Compulsory heterosexuality' represents other sexualities as marginal and deviant and, in doing so, reaffirms its own centrality and dominance. Two important tasks in the study of sexuality are, first, to make other sexualities visible and to resist their marginalisation; and, second, to destabilise the naturalisation of heterosexuality by asking why and how such a process of naturalisation occurred and by examining the sexed nature of heterosexuality. In other words, although geographies of sexuality may focus to a great extent on the lives of lesbians and gay men (and to a far lesser extent on the lives of bisexuals) in an attempt to make such lives visible and to challenge 'compulsory heterosexuality', geographies of sexuality should also address heterosexual relations between men and women and the ways in which such relations have become dominant over space and time.

Sexuality, power and knowledge

The term 'homosexuality' is a recent one, first entering public discourse in 1869 in the writings of the Hungarian journalist, Benkert von Kertbeny (Weeks, 1991). The term was not widely used in English until the 1880s and 1890s. At this time, homosexuality was identified as a medical problem that required treatment and as a legal problem that required legislation. But same-sex desire was clearly nothing new in the late nineteenth century, and was – and is – clearly not restricted to western cultures, as illustrated in Box 4.2. Recognising the need for historical and geographical specificity, this chapter will concentrate on dissident sexualities in Europe and North America from the late nineteenth century to the present.

Box 4.2 Same-sex desire over space and time

Same-sex love is a phenomenon common to almost every culture, one occurring throughout recorded history. The ways in which people have understood this attraction, however, have varied widely. For some cultures, such a love is natural and desirable. In ancient Greece, love for boys was seen as evidence of virility, and the relation between boys and men was crucial for the development of boys into men. The distinction was not between homosexual and heterosexual, but between passive and active; while boys could be the object of male affection and desire, the beloved, as they grew into men they were required to assume the posture of the lover instead. We see such distinctions today in many Islamic and Latin American societies. In Islamic Africa, for example, both men and women may have same-sex relationships, but these relationships are typically between wealthy older patrons and poorer, younger companions.

In some places and times, the attraction to another man or another woman has been interpreted as evidence that the person is not really a man or a woman, but is a hybrid or placed inside that body; the North American *berdache* may be the most prominent example. In Hawaii, *moe-aikane* (men who have sex with other men) and *mahus* (men who dress as women) are largely accepted in their families and social groups.

In other societies, most notably Jewish and modern Christian-dominated ones, love between men or between women is prohibited, and those who persist in it are stigmatized. This fear and hatred has provided much of the context for European and North American lesbian and gay politics. (Blasius and Phelan, 1997: 2)

Dissident sexualities became increasingly individualised over the course of the nineteenth century in Europe and North America. While dissident sex acts had previously been thought of in terms of sin, they now came to be legislated against and punished as crimes. Moreover, rather than concentrate on categorising sex *acts*, by the nineteenth century, *individuals* came to be categorised as homosexual for the first time. In many modern capitalist societies, legislation came to criminalise individuals as homosexual by the late nineteenth century. For example, in Britain, an act of Henry VIII in 1533 legislated against sodomy, and remained the basis for all homosexual convictions until 1885. But, as Jeffrey Weeks explains,

> the central point was that the law was directed against a series of sexual acts, not a particular type of person. There was no concept of the homosexual in law, and homosexuality was regarded not as a particular attribute of a certain type of person but as a potential in all sinful creatures. (Weeks, 1977: 12)

The 1533 Act legislated that the 'Abominable Vice of Buggery' should be punished by death. While the death penalty for buggery was largely abandoned after 1836, it remained on the statute books until 1861 in England and Wales and until 1889 in Scotland, when it was replaced by penal servitude of 10 years to life. As Weeks points out, abolishing the death penalty for buggery 'was a prelude not to a liberalization of the law but to a tightening of its grip' (Weeks, 1977: 14), in the form of the Labouchère Amendment to the Criminal Law Amendment Act, which was passed in 1885 (see Box 4.3). This amendment made *all* male homosexual acts illegal and criminalised male same-sex acts in all spaces, both public and private. This remained the case in England and Wales until as recently as 1967. Gay sex remained illegal in Scotland until 1981 and in Northern Ireland until 1982. Three and a half centuries after the 1533 Act, the Labouchère Amendment focused on the 'male person' committing certain sexual acts rather than purely on the acts themselves. Rather than represent such acts as sins to which sinful men might succumb, legislation in the late nineteenth century served to criminalise male homosexuals. Such legislation was explicitly gendered. While women were subject to laws concerning marriage, property rights and the notorious Contagious Diseases Acts of the 1860s, same-sex acts between women were not criminalised by the Labouchère Amendment or by other laws. In 1921, an attempt to broaden the 1885 provisions to include women failed, in part 'on the grounds that publicity would only serve to make more women aware of homosexuality' (Weeks, 1991: 19). Although lesbians were not criminalised in legal terms like homosexual men, they still suffered discrimination and prejudice in many spheres of life.

Box 4.3 Sex laws in the United Kingdom

The Labouchère Amendment of the 1885 Criminal Law Amendment Act made *all* male homosexual acts illegal in the UK:

> *Any male person who, in public or private, commits, or is a party to the commission of, or procures or attempts to procure the commission by any male person of any act of gross indecency with another male person, shall be guilty of a misdemeanour, and being convicted thereof shall be liable at the discretion of the court to be imprisoned for any term not exceeding two years, with or without hard labour.* (Quoted in Weeks, 1977: 14)

Box 4.3 *continued*

The Labouchère Amendment remained on the statute books until 1967, when the Sexual Offences Act (or Wolfenden Act) decriminalised *some* gay sex in explicitly spatial terms. This Act legislated that male gay sex in England and Wales was lawful only in private, with one partner, and for men aged over 21 (Binnie, 1995). Although the age of consent for gay men was lowered to 18 in 1994, it still remains two years higher than its heterosexual equivalent, although this law is likely to change in the near future. In the 15 countries of the European Union, only Britain, Finland and Austria have an unequal age of consent for homosexuals and heterosexuals.

By the mid-1980s, Britain 'had one of the most restrictive sets of legislation on gay male sexuality in Europe' (Binnie, 1995: 188), perhaps epitomised by Section 28 of the 1988 Local Government Act. Section 28 prohibits the 'promotion of homosexuality' by local governments and teaching about homosexuality as a 'pretended family relationship'. Although no prosecutions have been brought under Section 28, it continues to represent the heterosexual nuclear family as the norm against which other sexualities are seen to deviate. Other British legislation discriminates against gay men and lesbians in terms of immigration, workplace benefits, and the ability to foster or adopt a child.

The 1880s and 1890s have been represented as a time of 'sexual anarchy' (Showalter, 1990), when established gender roles were thought to be breaking down and a number of sexual scandals made the public more aware of sexual dissidence. Purity campaigns, often invoking racialised discourses of degeneration, sought to reaffirm the importance of the family at a time when it was thought to be threatened by sexual decadence. Legislation against male same-sex relationships led to a number of infamous trials that were well publicised in the popular press. The most sensational of these was the trial and conviction of Oscar Wilde in 1895. The famous wit and writer was accused of 'posing as a sodomite' by the father of his lover, Lord Alfred Douglas. Wilde sued for libel but the trial resulted in his eventual prosecution and conviction. He was found guilty of buggery and, in line with the Labouchère Amendment, was sentenced to two years in prison with hard labour. Passing judgement on Wilde and Alfred Taylor, who was convicted at the same time, Mr Justice Wills said that

> the crime of which you have been convicted is so bad that one has to put stern restraint upon one's self to prevent one's self from describing, in language which I would rather not use, the sentiments which must rise to the breast of every man of honour who has heard the details of these two terrible trials. ... It is no use for me to address you. People who can do these things must be dead to all sense of shame, and one cannot hope to produce any effect upon them. It is the worst case I have ever tried. (Quoted in Blasius and Phelan, 1997: 113)

The trial of Oscar Wilde 'was a vital moment in the creation of a male homo-sexual identity' (Weeks, 1991: 19). For the first time, the trial 'gave a public face to homosexuality, albeit one of negative stereotype and pathological deviance' (Blasius and Phelan, 1997: 111).

Sexuality came to be increasingly regulated in the late nineteenth century. At the same time, sexuality attracted an increasing amount of scientific attention, particularly in Germany and Britain. Just as male homosexuals came to be named, labelled and categorised as different and often deviant from heterosexuals, sexuality itself came to be named, labelled and categorised as an object of study. A great deal of early work on sexuality focused on its biological nature and basis. In more recent years, the biological determinism of much of this work has been criticised by writers who stress the social production of sexual relations and knowledges about sexuality, the fluidity within and between different sexualities, and the political significance of dissident sexualities. More recent and critical ideas about sexuality have largely been inspired by the psychoanalysis of Sigmund Freud and Jacques Lacan, which explores sexuality in the realm of the unconscious, and by the discourse analysis of Michel Foucault, which locates sexuality in a historically and geographically specific nexus of power/knowledge. Increasingly critical histories of sexuality have been written that reveal the links between sexuality, power and knowledge (see, for example, Duberman *et al.*, 1989; Milligan, 1993; Weeks, 1977, 1991). The rest of this section will explore early studies of sexuality in Germany and Britain and the ways in which such studies were explicitly tied to sexual politics. It will end by introducing the influential work of Freud, Lacan and Foucault.

Early sexology: biology and nature

Early sexology – the study of sex – was concerned with discovering and explaining the biological basis of sexual behaviour. The work of writers such as Karl Heinrich Ulrichs and Magnus Hirschfield points to the importance of German sexology, while the work of writers such as Havelock Ellis and Edward Carpenter reflected the growth and influence of sexology in Britain. All of these writers were studying sex in the wake of Charles Darwin's landmark analysis of evolution. As Jeffrey Weeks explains, Darwinism 'led to a revival of interest in the sexual "origins" of individual behaviour, and a sustained effort to delineate the dynamics of sexual selection, the sexual impulse, and the differences between the sexes' (Weeks, 1991: 70). Biological and psychological studies came to be privileged in this field as they sought to uncover the basis and manifestation of human sexuality in scientific terms.

In Germany, Paragraph 175 of the German Imperial Penal Code had been in place since 1871 (see Box 4.4). An early sexologist, Karl Heinrich Ulrichs (1825–1895), campaigned unsuccessfully to prevent its ratification, claiming that men who felt same-sex desire were not 'actual men' and, because they were subject to another 'law of nature', they were incapable of changing their behaviour (Ulrichs, in Blasius and Phelan, 1997: 64).

Box 4.4 Paragraph 175 of the German Imperial Penal Code, 1871

Like the later Labouchère Amendment in the UK, Paragraph 175 legislated against same-sex relationships between men:

> *Unnatural vice committed by two persons of the male sex or by people with animals is to be punished by imprisonment; the verdict may also include the loss of civil rights.* (Quoted in Blasius and Phelan, 1997: 63)

Alongside bestiality, homosexual relations were criminalised as an 'unnatural vice' that had to be punished. As in the UK, same-sex relationships between women were not included in this sexual legislation. Paragraph 175 was tightened when Hitler came to power in 1933. The Nazi version remained law in West Germany until it was reformed in 1969. In East Germany, it was struck down in 1948, Paragraph 175 was reformed in 1968 and was repealed in 1988. German unification in 1990 meant that West German law came to apply to the former East Germany. After a two-year moratorium, Paragraph 175 was finally repealed in 1994 (Blasius and Phelan, 1997: 134).

Ulrichs published 12 books on homosexuality between 1864 and 1879 and is perhaps best known for analysing same-sex desire as a 'third sex' or an 'intermediate sex'. Ulrichs described men who desired other men as 'Urnings' or 'Uranians'; lesbians as 'Uraniads'; people who desired the opposite sex as 'Dionings'; and people who desired both men and women as 'Uranodionings'. These terms were inspired by Plato's *Symposium*, in which he described two types of love, each ruled by a goddess called Aphrodite, one of whom was the daughter of Uranus and one the daughter of Zeus and Dione (Blasius and Phelan, 1997). For Ulrichs, men and women who desired those of their own sex had hermaphroditic personalities, mixing both male and female characteristics. As he wrote, 'the urning is not a true man. He is a mixture of man and woman. He is man only in terms of body build. The love drive inherent to him, on the other hand, is that of a female being. Accordingly, it must be directed toward the male sex' (Ulrichs, in Blasius and Phelan, 1997: 64). Ulrichs theorised the 'third sex' of same-sex desire in terms that reproduced heterosexual gender relations *between* the sexes. So, for example, he wrote that 'urnings' were men who embodied *female* desires for other men, assimilating homosexual desire into a heterosexual structure.

By the 1890s, there was an unprecedented proliferation of work on sexuality in Germany and, by the beginning of the twentieth century and until the early 1930s, Germany was home to the most thriving gay and lesbian political movement in the world (Box 4.5). By the early twentieth century, 'Germany became an international beacon for lesbian and gay political culture much as the U.S. is today' (Blasius and Phelan, 1997: 134). In 1897, the Scientific-Humanitarian Committee was founded by Magnus Hirschfield, aiming to abolish Paragraph 175 of the German Imperial Penal Code by conducting scientific research that would change sexual attitudes and prejudices. This Committee was the first and the most significant of early gay rights organisations, and explicitly tied the production of knowledge about sexuality to sexual politics. The first two clauses of the Committee's constitution detailed its scientific, educational and political aims:

1. The *aim* of the Committee is research into homosexuality and allied variations, in their biological, medical and ethnological significance as well as their legal, ethical and humanitarian situation. The S.H.C. gives its members every assistance in the spirit of true humanitarianism.

2. The Committee wants to change public opinion about homosexuality through publications ..., pamphlets and petitions, scientific talks and popular lectures. (Quoted in Wolff, 1986: 449)

Box 4.5 Gay and lesbian politics in Germany in the early twentieth century

By the turn of the century ... and during the entire first third of the twentieth century, Germany differed from all other countries by virtue of its gay and lesbian political movement and comprehensive cultural formation.

Why Germany? To begin with, Germany's status as a 'belated' nation in comparison with other Western European countries evoked a compensatory eagerness to embrace modernization, at least on the part of bourgeois liberals. The tempo of transformation invigorated both the nation's intellectual elite and its powerful socialist movement, who saw themselves as the cutting edge of social and political progress; and even the conservative imperial regime was likewise determined to propel Germany to world leadership in commerce and culture. Dire social dislocation resulting from rapid urbanization spawned an innovative middle-class 'life reform movement,' which provided an oppositional context in which the homosexual emancipation movement could crystallize. In addition, the uneven but largely successful course of Jewish emancipation in the nineteenth century served as a model for the acceptance of other minorities, and a notable number of gay leaders were of Jewish ancestry. Finally, in an increasingly secular age, Germany attached exceptional prestige to science, which enjoyed the special patronage of Kaiser Wilhelm II; striking advances were made in the natural and social sciences. (Blasius and Phelan, 1997: 133)

On behalf of the Scientific-Humanitarian Committee, Hirschfeld drew up the first of several petitions to the German Reichstag or parliament for the abolition of Paragraph 175 in 1897. This petition argued that 'the abolition of similar punishments in France, Italy, Holland and many other countries had never lowered moral standards' and that 'the real cause of homosexuality ... is due to development of the bisexual nature of man. The human foetus, during its first three months, is a bisexual organism. Therefore no moral guilt can possibly be attributed to homosexual sentiments' (Hirschfeld, in Blasius and Phelan, 1997: 135-136).

Magnus Hirschfeld (1874–1945) was a Jewish physician who, through his biological research, sought to encourage both homosexuals and wider society to regard same-sex desire as natural. His famous slogan, 'Through knowledge to justice', represented his commitment to scientific research, public education and political reform (Weeks, 1977). In his first publication on sexuality, Hirschfield commented on the trial of Oscar Wilde, which had attracted a great deal of publicity and interest in Germany. As he wrote, 'The married man who seduces the governess of his children remains free, as free as the countess who has a liaison with her maid. But Oscar Wilde, this genius of a writer, who loves Lord Alfred Douglas with a passionate love, has been put into prison in Wandsworth. And this because of a passion which he shares with Socrates, Michelangelo and Shakespeare' (quoted in Wolff, 1986: 33). Hirschfield developed a typology of sexual desire, classifying different sexual orientations according to foetal development and hormonal influences. Although this biological basis led Hirschfield to recognise same-sex desire as *natural*, he referred to opposite-sex desire as *normal*. For example, he compared same-sex desire with biological differences such as hare-lip or club-foot, which he saw as abnormal even though he regarded them as natural and biologically based (Wolff, 1986). Hirschfield's extensive research and writings culminated in the encyclopaedic publication in 1914 of *The Homosexuality of Men and Women*, which drew on the histories of 10,000 homosexual men and women and extended to more than 1000 pages in length (Wolff, 1986). In this and his other work, Hirschfield provided 'the definitive statements of the biological–congenital model of homosexuality' (Weeks, 1977: 129), describing and explaining different sexualities in terms of biological factors underpinning an essential human 'nature'. In 1919, Hirschfield opened the first Institute for Sexual Science in the world. The Institute in Berlin was a centre for research, teaching and consultations for people with sexual difficulties, 'be they due to marital problems or unorthodox love' (Wolff, 1986: 175).

In Britain, writers such as Havelock Ellis were influenced by the work of Ulrichs and Hirschfield. Havelock Ellis (1859–1939) wrote about sexual desire as a positive and life-enhancing force. Like Ulrichs and Hirschfield, he located sexuality firmly within a biological rather than a social realm, attempting to explain human behaviour in terms of natural impulses that could not and should not be suppressed. According to Cate Haste, 'he acknowledged the centrality of the sexual impulse as a motivating factor in human development, recognized the importance of infant sexuality, and concluded that it was the denial and repression of the sexual impulse, not its expression, which wrought the greater damage to the development of an essentially healthy drive' (Haste, 1994: 22). Ellis wrote about the same-sex desire of some people as 'sexual inversion', representing it as a congenital condition. Ellis argued that while some people could choose to participate in sexual relations with someone of their own sex, or could be corrupted into such relations, others were born as sexual inverts, *naturally* desiring others of the same sex. As Jeffrey Weeks shows, Ellis's ideas were highly influential in campaigns for sexual reform and in many ways continue to provide the basis of liberal understandings and 'tolerance' of homosexuality today: 'Three elements have been central: first, the argument that homosexuality is characteristic of a fixed minority and incurable; secondly, that reforming efforts should be directed towards changing the law so that this minority may live in peace; and thirdly, the belief that such

reform would only come about through a long period of public education'
(Weeks, 1977: 64–65). As Weeks argues, the underlying assumption remains
'that homosexual behaviour has to be explained as a deviation from a norm of
sexual behaviour' (Weeks, 1977: 65).

The first International Congress for Sex Reform was convened by Hirschfield in Berlin in 1921. At a subsequent congress, held in 1928 in Copenha-
gen, Hirschfield and Ellis were elected as two of the founding presidents of the
World League for Sexual Reform. According to Weeks, the League aimed 'to
harmonize social and judicial practice with the "laws of nature"':

> Its specific planks included support for the political, economic and sexual
> equality of women and men; reform of marriage and divorce laws; improved
> sex education; the control of conception; reform of the abortion laws; the
> prevention of venereal disease and prostitution; and the protection of
> unmarried mothers and the illegitimate child. (Weeks, 1977: 139)

By stressing the 'laws of nature' in line with biological conceptions of same-sex
desire, the League also addressed the reform of laws regulating homosexuality,
arguing for 'a rational attitude towards sexually abnormal persons, and especially towards homosexuals, both male and female' (quoted in Weeks, 1977:
139).

Sexuality was the subject of increasing scientific attention in Europe from
the late nineteenth century to the early 1930s. Despite prejudice, legal restrictions and censorship, a gay and lesbian political movement began to emerge
that was often closely linked to feminist and socialist movements. In the early
twentieth century, Germany was the main centre for lesbian and gay research
and politics, as shown by the establishment of the Scientific-Humanitarian
Committee, the Institute of Sexual Science, and the political campaign organised by Ulrichs and Hirschfield. In Berlin alone, the publication of 30 lesbian
and gay periodicals reflected a burgeoning subculture (Blasius and Phelan,
1997). In the late 1920s, the World League for Sexual Reform represented an
international attempt to campaign for sexual reform. In both cases, the production of scientific knowledge about sexuality was closely tied to sexual politics. At this time, most ideas about human sexuality continued to search for
sexual origins in biological terms, suggesting that sexual orientations were
innate rather than socially produced. While homosexuality was seen as natural
because it was determined by biology, it was seen as an abnormal deviation
from heterosexuality. Identifying homosexual men and women as an intermediate 'third sex' continued to define same-sex desire in terms of heterosexual
gender relations.

The proliferation of sexual research and sexual politics came to an abrupt
end when Hitler came to power in Germany in 1933, tightened Paragraph 175
and banned all public manifestations of homosexuality. After giving a public
lecture, Hirschfield was brutally attacked and his skull was fractured (Haeberle, 1989). As a Jew, a homosexual and a sexologist, Hirschfield was a clear
target for Nazi persecution. He fled to Switzerland and then to France, where
he died in exile in 1935. In May 1933, the Institute for Sexual Science was

raided and its library, archives, the records of the World League for Sexual Reform and a bronze bust of Hirschfield were all burnt in a public fire. In the autumn of 1933, the first homosexuals were sent to concentration camps. By 1945, the Nazis had arrested 90,000 homosexuals and bisexuals, and had sent 10,000 to concentration camps, where they were identified by pink triangles on their uniforms (Blasius and Phelan, 1997; also see Haeberle, 1989; Plant, 1988). Their mortality rate was very high because of murder, torture and dangerous work. As Haeberle writes, 'The verbal denigration of homosexuals, their stigmatization, imprisonment, and finally, forced "cures" for their alleged medical condition – in all these respects the Nazis merely continued and intensified what had long been general practice and what, in various forms, still continues in many societies, including our own' (Haeberle, 1989: 378).

Spaces of the psyche: Freud and Lacan

From the beginning of the twentieth century, the psychoanalytic theories developed by Sigmund Freud inspired new ways of thinking about sexuality. Rather than concentrate on uncovering a biological essence of human nature, Freud (1856–1939) focused on 'an unknown, unexplored and indefinite region between the surface of the mind and the body' (Milligan, 1993: 15). He termed this region the *unconscious*, which existed separately from, but in creative tension with, biological 'nature' and human society. Through his clinical practice as a consultant on nervous disorders in Vienna, Freud developed a psychoanalytic approach that located sexuality in the realm of the unconscious and analysed sexual desires in the workings of the mind. In many ways, Freud's theories of the unconscious were inherently spatial, invoking ideas about location, distance, proximity and the gaze (see Bondi, in McDowell and Sharp, 1999, for a helpful summary of Freudian theory).

In his clinical practice, Freud moved away from traditional treatments such as hypnosis and electrotherapy to develop a 'talking cure' based on free association (the patient saying whatever was in their mind) and the interpretation of dreams. Although free association often resulted in an apparently random, disconnected stream of consciousness, Freud believed that it was all related to the subject under discussion. But, as Milligan explains, 'he had to account for the fact that any rational connection between the patient's talk and the particular matter in hand often appeared to be absent. He had to account for connections that were unrecognisable. This he did by postulating the existence of *resistances* that obstructed the patients' access to their own unconscious' (Milligan, 1993: 18). In other words, Freud argued that the unconscious served to *repress* certain feelings and desires. The task of the analyst, according to Freud, was to expose and to overcome such resistances and repressions. As Liz Bondi explains, 'many of the wishes and desires that constitute the repressed unconscious relate to sexuality' (Bondi, in McDowell and Sharp, 1999: 98).

Two central elements of Freudian psychoanalysis are the Oedipus complex and the castration complex, which are seen to be experienced differently by boys and girls. In both cases, Freud argued that humans are born bisexual but that as they develop, their sexual identities and gender roles become intimately bound up together. Unlike the work of Ulrichs and Hirschfield that identified same-sex desire as a 'third sex' or an 'intermediate sex', Freud stressed that same-sex and opposite-sex desires are both important parts of the human unconscious. Freud traced the psychic significance of biological difference, whereby boys and girls experience their sexuality differently and ultimately grow up as distinctly gendered as well as sexed subjects. As Rosemary Tong puts it, 'If men adjust to their sexual maturation normally (that is, typically), they will end up displaying expected masculine traits; if women develop normally, they will end up displaying expected feminine traits' (Tong, 1992: 139–140). For Freud, such 'normal' psychic and sexual development resulted in the opposite-sex desire between men and women.

Unlike the biological studies of Ulrichs, Hirschfield and Ellis, Freud examined the unconscious as the main site for the articulation and repression of sexualities. But his views of the unconscious were intimately bound up with ideas about biological difference. In contrast, the influential work of Jacques Lacan (1901–1981) explored the unconscious in terms of language as well as biology (see Bondi, in McDowell and Sharp, 1999, for a helpful summary of Lacanian theory). Lacan analysed the development of a speaking subject moving away from the imaginary realm of the unconscious as a site of repressed meanings to the symbolic order of language, laws, institutions and social processes. For Lacan, the mirror stage represents the key stage when a child first perceives itself as an individual, existing separately from its mother. This self-identification is underpinned by viewing the presence or lack of a phallus. As a result, the phallus becomes the signifier of sexual difference, gender identity, and sexual desire between men and women. Once again, sexualities are inherently spatial, depending on proximity, distance and the gaze.

The psychoanalytic theories of Freud and Lacan have been extremely influential, moving away from purely biological studies that sought to uncover the essence of human nature to examine the complex workings and repressions of the unconscious. Rather than view sexuality as biologically determined, both Freud and Lacan saw sexuality as relational, with boys and girls coming to identify themselves as sexed and gendered subjects in different ways. But in many ways, such ideas continue to posit heterosexuality as the norm against which other sexualities are represented as deviant. Furthermore, female sexuality is interpreted as 'lack' in terms of its difference and supposed incompleteness in contrast to male sexuality. Many critiques of the work of Freud and Lacan argue that it remains both heterosexist and patriarchal. And yet, many feminist theorists have been inspired by such work. Feminists such as Hélène Cixous, Luce Irigary and Julia Kristeva have developed Lacan's reworkings of Freudian psychoanalysis in complex and critical ways (see Tong, 1992, and Weedon, 1997, for introductions to their work). Moreover, Gillian Rose has built on Lacanian psychoanalysis in her critique of the phallocentric masculinism of a distanced, objectifying geographical gaze on landscape, as discussed in Chapter 3 (Rose, 1993).

The work of both Freud and Lacan explores sexuality in spatial terms, positioning subjects in proximity and at a distance from each other and examining the importance of the gaze (see Chapter 3 for further discussion of the geographies of a gendered gaze). But at the same time, the work of both Freud and Lacan has been criticised for being aspatial and ahistorical as it examines the psychic constitution of gender and sexuality in universal terms, removed from the specificity of material contexts over space and time. As Chris Weedon says, 'Psychoanalysis offers a universal theory of the psychic construction of gender identity on the basis of repression' (Weedon, 1997: 43). The construction of gender identity is inextricably bound up with sexual identities in a heterosexist and patriarchal system. While such a universal theory might reflect and attempt to explain the dominance of heterosexuality, it also appears to be abstracted and decontextualised. In contrast, the work of Michel Foucault explores sexuality as discourse, located within historically and geographically specific contexts of power/knowledge.

Sexuality as discourse: Foucault

For Michel Foucault, sexuality is discursive, material and inseparable from the exercise of power and the associated production of knowledge (Foucault, 1990; and see Chapter 3 for more on discourse). So, for example, 'compulsory heterosexuality' is discursively produced through the media, law, religion and education (see, for example, Haste, 1994, for an account of the 'rules of desire' that have influenced sexualities in Britain since the First World War). Foucault has written about the discursive production of sexuality, the ways in which such discursive formations vary over space and time, and the ways in which certain discursive practices become institutionalised (also see Foucault's work on the discursive production of madness: Foucault, 1973). Discourses, and the material inscription of discursive practices, are inseparably bound up with power relations, which, in poststructuralist terms, are diffused throughout society (see Chapter 3 for an introduction to poststructural feminism).

First published in 1976, the first volume of Foucault's *The History of Sexuality* has been particularly influential in challenging conventional histories of sexuality. Foucault represents such histories as contrasting a repressed nineteenth-century sexuality with its modern, more liberated counterpart. Foucault challenges Freudian notions of repression by critiquing what he terms the *repressive hypothesis*, which refers to the repression of desire in the context of the private sphere. Foucault also critiques the sexual *essentialism* that characterises many histories of sexuality, whereby sexuality is identified as a separate, repressed domain of human existence invested with the true essence of human nature and identity. In contrast, Foucault examines the productive mechanisms through which power over sexuality is exercised and argues that sexuality is *constructed* by the very discourses that claim to be describing and analysing it as an object of study. As Nancy Wood explains, in the nineteenth century

'sexuality' was in fact *constructed* by all those discourses – religious, legal, psychiatric, educational, medical – which claimed merely to *describe* it. Foucault's proposed *history* of sexuality, then, is guided by the aim of analysing, not an essential sexuality, but *what* various discourses said about sexuality, *why* they emerged when they did, and some of the consequences of these pronouncements.
(Wood, 1985: 162)

For Foucault, sexual discourses are inseparable from the historically and geographically specific configurations of power/knowledge. He explores the ways in which the exercise of power and the production of knowledge are inextricably bound up together through, for example, the administration, regulation and control of population growth and through the emergence of *scientia sexualis* or science of sex, encompassing medicine, psychiatry, sexology and psychoanalysis, all of which make sexuality the object of scientific knowledge. Foucault critiques the essentialist notion that it is only possible to uncover the 'truth' about an individual by uncovering the 'secrets' of their repressed sexuality. In this way, Foucault represents the nineteenth-century science of sex as a secular counterpart to a Catholic confession. Both operate by positing a hidden, essential and 'true' meaning that is repressed and needs to be revealed. Foucault argues that rather than revealing 'true' and essential meanings about sexuality, such analyses act as techniques of power, and that such techniques of power come to be internalised by people who regulate and discipline their own behaviour accordingly.

Unlike work on the biological and psychoanalytic nature of sexuality, Foucault explores the links between sexuality, knowledge and power, and traces the ways in which these links have changed over space and time. He examines the ways in which controls over sexuality are exercised through 'regulative' measures that are represented in negative terms as, for example, repression, prohibition and censorship. But Foucault argues that these controls are *productive* rather than repressive, producing different discourses of sexuality rather than repressing its fixed and essential meaning. For Foucault, disciplinary power operates through networks on all levels in society, subjecting individual bodies as well as the social body through self-surveillance and self-discipline.

While the early work of sexologists such as Ulrichs, Hirschfield and Ellis explicitly tied the production of knowledge about sexuality to sexual politics, the work of Foucault examines the discursive constitution of such links between sexuality, knowledge and power. This section has introduced some of the ways in which ideas about sexuality are diverse and contested, spanning biological, psychoanalytic and discourse analyses. The work of Freud, Lacan and Foucault represents sexuality in explicitly spatial terms and has been particularly influential in geographical analyses of sexuality. Before exploring the ways in which geographers have studied sexuality over the course of the 1980s and 1990s, it is important to contextualise this work in terms of post-war gay liberation movements.

Gay liberation

While Germany was home to the most thriving lesbian and gay political culture in the world in the early twentieth century, the same can be said of the United States since the mid-twentieth century. According to Blasius and Phelan, 'The United States was something of a latecomer to gay and lesbian activism' (Blasius and Phelan, 1997: 217). The Stonewall riot of 1969 is often identified as the beginning of dissident sexual politics in the United States and is discussed below. But the roots of American gay liberation were laid much earlier:

> A mythology has grown around the Stonewall riots. In this myth, Stonewall was the beginning of gay and lesbian political activism and culture. In fact, the groundwork for that historical event was laid by decades of hard work ... By the time Stonewall occurred, the groundwork had been laid for a full political and cultural movement. (Blasius and Phelan, 1997: 239)

The Chicago Society for Human Rights, founded in 1924, was the first documented gay organisation in the United States. A number of its members were arrested and prosecuted for founding a 'strange sex cult' that 'urged men to leave their wives and children' (quoted in Blasius and Phelan, 1997: 217) and, although these members were released due to a legal technicality, they continued to suffer harassment and the society was disbanded in 1925. In New York, the so-called Harlem Renaissance lasted from about 1920 to 1935 and represented a key moment not only in African–American history but also in the history of gay and lesbian Americans. The growth of industrial employment in northern cities, exacerbated by the United States fighting in the First World War, led to the large-scale migration of black Americans from the rural south to the urban north in the first two decades of the twentieth century: 'Black communities developed in Chicago, Detroit, and Buffalo, but the largest and most spectacular was Harlem, which became the mecca for Afro-Americans from all over the world. Nowhere else could you find a geographic area so large, so concentrated, really a city within a city, entirely populated by blacks' (Garber, 1989: 319). In the 1920s, Harlem became famous for its literature, blues, jazz and clubs: 'Calling themselves the "New Negroes", residents expected and demanded full and equal participation in American cultural, social, and political life. Homosexuality was integral to, and to some degree an accepted part of, the new cultural life of Harlem' (Blasius and Phelan, 1997: 227). A homosexual subculture in Harlem emerged in the context of oppression and poverty: 'in spite of racial oppression, economic hardship, and homophobic persecution, black lesbians and gay men were able to build a thriving community of their own within existing Afro-American institutions and traditions' (Garber, 1989: 321). Harlem's speakeasies, private parties and drag balls helped to create a social network for lesbians and gay men, which continued to thrive until the Stock Market crash of 1929, the subsequent Depression, and the repeal of Prohibition (see Figure 4.3).

Figure 4.3 The Harlem Renaissance – entertainers at Small's Paradise Club

According to John D'Emilio, 'The slow, gradual evolution of a gay identity and of urban gay subcultures was immeasurably hastened by the intervention of World War II ... [The war] was something of a nationwide coming-out experience. It properly marks the beginning of the nation's ... modern gay history' (D'Emilio, 1989: 458). The social disruption of war, when millions of American men and women moved away from their homes and families to live in non-familial, same-sex environments for the first time, 'created a setting in which to experience same-sex love, affection, and sexuality, and to discover and partici-pate in the group life of gay men and women' (D'Emilio, 1989: 459). Focusing on the development of gay politics and a gay community in San Francisco, John D'Emilio points to the decisive impact of the Second World War. San Francisco was a major port for American servicemen and women travelling to and from the war in the Pacific and was an important centre for war industries. While the population of the city had declined in the 1930s, it increased by over 125,000 between 1940 and 1950, with many Americans remaining after demobilisation from the war (D'Emilio, 1989). As D'Emilio explains,

> The growth included a disproportionate number of lesbians and gay men. The sporadic, unpredictable purges from the armed forces in the Pacific deposited lesbians and homosexuals, sometimes hundreds at a time, in San Francisco with dishonorable discharges [sodomy became a crime in the US military in 1919 and homosexuals were officially banned in 1943]. Unable or unwilling to return home in disgrace to family and friends, they stayed to carve out a new gay life. California, moreover, was the one state whose courts upheld the right of homosexuals to congregate in bars and other public establishments. Though the police found ways around the decision and continued to harass gay bars, the ruling gave to bars in San Francisco a tiny measure of security lacking elsewhere. By the late 1950s about thirty gay male and lesbian bars existed in the city. (D'Emilio, 1989: 459)

The harassment and persecution of lesbians and gay men in the United States plumbed new depths in the post-war McCarthy era. In the 1950s, representations of the nuclear family and traditional gender roles and sexual behaviour were crucially important to Cold War politics. The McCarthyite persecution of suspected communists and 'subversives' equated 'perversion' with treason, claiming that homosexuals would be liable to blackmail and would thus be likely to betray their country. As D'Emilio explains,

> If the war years had allowed large numbers of lesbians and gay men to discover their sexuality and each other, repression in the postwar decade heightened consciousness of belonging to a group. ... Women experienced intense pressure to leave the labor force and return home to the role of wife and mother. Homosexuals and lesbians found themselves under virulent attack: purges from the armed forces; congressional investigations into government employment of 'perverts'; disbarment from federal jobs; widespread FBI surveillance; state sexual psychopath laws; stepped-up harassment from urban police forces; and inflammatory headlines warning readers of the sex 'deviates' in their midst. The tightening web of oppression in McCarthy's America helped to create the minority it was meant to isolate. (D'Emilio, 1989: 459)

In 1952, the American Psychiatric Association formally classified homosexuality as an illness (Abelove *et al.*, 1993; this classification remained in place until 1973) and, in 1957, the American Civil Liberties Union declared that laws against consensual gay and lesbian sex and entrapment procedures used to arrest gay men did not fall within its purview (Blasius and Phelan, 1997). The 1950s was a decade of renewed conservatism in the United States. Same-sex desire was represented as a sin, a crime or an illness in the context of social, sexual and political conservatism, oppression and prejudice.

Despite, and in part because of, such conservatism, the 1950s witnessed the emergence of the first sustained gay and lesbian organisations in the United States. The activism of these 'homophile' organisations laid the foundations for post-Stonewall sexual politics from 1969. The largest of these organisations were based in California and included the Mattachine Society, the Daughters of Bilitis, and ONE (introduced in Box 4.6). By the late 1960s, regional chapters of the Mattachine Society and the Daughters of Bilitis established a political network in different places. These and other groups were brought together to discuss their aims and differences by the North American Conference of Homophile Organizations or NACHO. The meetings and publications of these organisations provided important channels for representing lesbian and gay issues and experiences. Unlike later political activism, these organisations generally agreed to work for the assimilation and integration of lesbians and gay men within heterosexual society, and did not challenge the heterosexist gender norms that underpinned social relations (Blasius and Phelan, 1997).

Box 4.6 Pre-Stonewall dissident sexual politics in the United States

The Mattachine Society was founded in Los Angeles in 1950. Its name comes from a medieval European term for 'societies of men who played the fool or the jester in dance, and in doing provided veiled political satire' (Blasius and Phelan, 1997: 283). In its original aims, the Society sought to foster a 'highly ethical homosexual culture ... paralleling the emerging cultures of our fellow minorities ... the Negro, Mexican, and Jewish Peoples' (quoted in Blasius and Phelan, 1997: 283). In 1953, the group was taken over by new leaders who moved away from this vision of a homosexual minority culture to pursue assimilation into wider heterosexual society. The overtly political activity of its early years came to be superseded by a concern with legal and scientific questions and research, attempting to locate the 'cause' of homosexuality and refusing a group identity as homosexual. Although less radical than its original form, this was 'the Mattachine that would open a space for (male) homosexuals to find one another and discuss their common problems' (Blasius and Phelan, 1997: 285).

ONE was the most radical of American homophile organisations in the 1950s, claiming that homosexuality was positive and life-affirming and describing a homosexual culture as different, more rebellious and more creative than the heterosexual mainstream. ONE magazine was first published in 1954 and outlined the group's aims: 'ONE does not claim that homosexuals are better or worse than anyone else, that they are special in any but one sense. And in that one sense ONE claims positively that homosexuals do not have the civil rights assured all other citizens. ONE is devoted to correcting this. ONE means to stimulate thought, criticism, research, literary and artistic production in an effort to bring the public to understand deviants and deviants to understand themselves as the two sides are brought together as one' (quoted in Blasius and Phelan, 1997: 309). The ONE Institute was founded in Los Angeles in 1952 and continues to operate today.

The Daughters of Bilitis was founded in San Francisco in 1955 and was the first lesbian organisation in the United States. Its name came from the title of a book of poetry that described the same-sex love of Sappho's school (Blasius and Phelan, 1997). From 1956, the DOB published a monthly magazine, *The Ladder*. In its statement of purpose in 1955, the DOB described itself as 'a women's organization for the purpose of promoting the integration of the homosexual into society' (reprinted in Blasius and Phelan, 1997: 328). Education and research were seen as central planks of such integration, educating both 'the variant' and 'the public at large' about homosexuality. The DOB also sought to advance the cause of integration 'by advocating a mode of behavior and dress acceptable to society' (quoted in Blasius and Phelan, 1997: 328).

At the end of the 1960s, after almost 20 years of homophile activity, there were fewer than 50 organisations in the United States. But by 1973, there were more than 800 lesbian and gay groups throughout the country (D'Emilio, 1989: 466). At the same time as the Civil Rights movement, women's liberation and anti-Vietnam War protests were radicalising politics and politicising questions of identity, the Stonewall riot and the formation of the Gay Liberation Front (GLF) resulted in 'the birth of contemporary lesbian and gay politics' (Blasius and Phelan, 1997: 377). Both the Stonewall riot and the formation of the GLF took place in 1969 in New York. Both inspired an unprecedented level of dissident sexual politics that spread throughout the United States and beyond to other western countries. The New York Mattachine Society called the Stonewall riot 'The hairpin drop heard all around the world' (Healey and Mason, 1994: 6), which revolutionised dissident sexual politics. The London GLF was formed in 1970 and the Front Homosexuel d'Action Révolutionnaire was formed in France in 1971. Similar movements were established in Italy, Germany, Belgium, Holland, Canada, Australia and New Zealand (Weeks, 1977).

On 28 June 1969, the Stonewall, a gay bar in New York's Greenwich Village, was raided by police. As Weeks puts it, 'This was a regular occurrence, but this time the reaction was different – the homosexuals fought back' (Weeks, 1977: 188; also see Duberman, 1994). The Stonewall riot lasted for five days: 'The result was a kind of liberation, as the gay brigade emerged from the bars, back rooms, and bedrooms of the Village and became street people' (quoted in Weeks, 1977: 188). For the first time, the battle for gay rights was taken to the streets and made public and visible. The GLF was formed in the wake of the Stonewall riot and transformed sexual politics from the integrationist strategies of homophile groups into public and highly visible freedom campaigns that asserted gay and lesbian identities. As the founding statement of the New York Gay Liberation Front proclaimed:

> We reject society's attempt to impose sexual roles and definitions of our nature. We are stepping out of these roles and simplistic myths. WE ARE GOING TO BE WHO WE ARE ... Babylon has forced us to commit ourselves to one thing – revolution. (quoted in Weeks, 1977: 188)

The GLF tied New Left politics to sexual politics in its opposition to capitalist and heterosexist oppression and sought liberation on both individual and societal levels. As D'Emilio suggests, the notion of 'coming out' served as a crucial goal and strategy as gay men and lesbians identified themselves in more public and visible ways. As he writes,

> Coming out became a profoundly political step that an individual could take. It promised an immediate improvement in one's life, a huge step forward in shedding the self-hatred and internalized oppression imposed by a homophobic society. Coming out also became the key strategy for building a mass movement. Gay women and men who came out crossed a critical dividing line. They relinquished invisibility, made themselves vulnerable to

attack, and became invested in the success of the movement. Visible lesbians and gay men, moreover, served as magnets that drew others to them. (D'Emilio, 1989: 466)

'Coming out' invokes an explicitly spatial image, as gay men and lesbians choose to move across a private, perhaps closeted, threshold into public acknowledgement of their sexuality (see Brown, 1999, for more on the spatial imagery of the 'closet'). Rather than attempt to assimilate into heterosexual society, the campaigns of the GLF represented the public and visible assertion of, and pride in, gay and lesbian identities *as* gay and lesbian. Just as the Stonewall riot was fought on the streets, the GLF sought to widen the terrain of sexual politics from the private to the public sphere, taking dissident sexualities 'Out of the closets, onto the streets!' (GLF slogan, quoted in Healey and Mason, 1994: 9). The first Gay Rights March was held in New York in 1970 and attracted 5000 women and men. This march was the predecessor of annual Gay Pride marches that reclaim public space for lesbians and gay men today in many cities such as London, New York and Sydney (see Figure 4.4).

Campaigns for gay liberation took place alongside campaigns for women's liberation, as discussed in Chapter 3. Post-Stonewall sexual politics were also characterised by the emergence of a lesbian movement that far exceeded the Daughters of Bilitis in scale and scope. Lesbians were actively involved in early gay liberation and feminist politics, but 'By 1970 the experience of sexism in

Figure 4.4 Stonewall 25 parade: gay pride, New York, 1994

gay liberation and of heterosexism in women's liberation inspired many lesbians to form organizations of their own, such as Radicalesbians in New York, the Furies Collective in Washington, D.C., and Gay Women's Liberation in San Francisco' (D'Emilio, 1989: 466–467). Lesbian–feminist writers such as Teresa de Lauretis, Audre Lorde and Adrienne Rich have been extremely influential in tracing the complex links between different identities and systems of oppression, along axes of power that include sexuality, gender, race, class and age (see, for example, de Lauretis, 1994; Lorde, 1984; Rich, 1986).

According to Jeffrey Weeks, 'The early 1970s mark the turning-point in the evolution of a homosexual consciousness. The homophile organizations that tiptoed through the liberal 1960s were superseded in the 1970s by a new type of movement which stressed openness, defiance, pride, identity – and, above all, self-activity' (Weeks, 1977: 185). The legacy of gay liberation movements from the early 1970s lives on today, in the work of political groups such as Stonewall and OutRage! in Britain, and Queer Nation and the National Gay and Lesbian Task Force in the United States (see Box 4.7 on Stonewall, OutRage! and Queer Nation).

Since the early 1980s, the emergence of HIV and AIDS has had a profound impact on dissident sexualities. While the HIV virus affects heterosexuals as well as homosexuals, representations of its threat and effects in the West have tended to focus on gay men. Over the course of the 1980s, the rise of New Right politics in Britain and the United States, and the rise of Christian fundamentalism, particularly in the United States, helped to create a moral climate that stigmatised gay men and their sex lives, and represented same-sex desire itself as a deadly disease (see Sontag, 1989, for an account of AIDS as metaphor, and see Showalter, 1990, for a discussion of a similar moral panic surrounding syphilis in the late nineteenth century). Weeks identifies three stages in the history of AIDS: the dawning crisis in 1981–82; a moral panic from 1982–85; and crisis management from 1985 onwards (Weeks, 1991), and describes the 1990s as an 'age of uncertainty' for gay men and lesbians (Weeks, 1995).

Box 4.7 Stonewall, OutRage! and Queer Nation

Stonewall was founded in 1989 in Britain and describes itself as 'the national civil rights group working for legal equality and social justice for lesbians, gay men and bisexuals'. Its current campaigns include the repeal of Section 28; an equal age of consent for heterosexuals and non-heterosexuals; the repeal of the offence of gross indecency between men; the extension of the Sex Discrimination Act to include sexual orientation; the end of the ban on lesbians and gay men in the armed forces; and the right to form a 'family of choice'. Another current campaign is called Equality 2000, which aims to achieve equality at school, at work, in love, as parents and as partners by the year 2000. Visit its website at *www.stonewall.org.uk* for more information.

OutRage! was founded in 1990 in Britain and 'is a broad based group of queers committed to radical, non-violent direct action and civil disobedience to:

Box 4.7 *continued*

assert the dignity and human rights of queers; to fight homophobia, discrimination and violence directed against us; and to affirm our right to sexual freedom, choice and self-determination'. OutRage! rejects what it regards as 'the assimilationist and conformist politics of the mainstream lesbian and gay movement'. It campaigns for an Unmarried Partners Act, to give legal rights to all unwed couples, both gay and straight; to lower the age of consent for all to 14; and for an Equal Rights Act 'that protects all citizens (including lesbians and gay men) against all forms of discrimination, harassment, and incitement to hatred'. Visit its website at *www.outrage.cygnet.co.uk* for more information.

Queer Nation was founded in 1990 in the United States and represents 'a loosely organised and ostensibly non-hierarchical political movement founded ... by members of the activist organisation AIDS Coalition to Unleash Power (ACT-UP) and others specifically to fight homophobia and heterosexism. The goal was to adapt "direct action" tactics and strategies which had proved successful in the political battle against HIV/AIDS to other issues, grounded in heterosexism and homophobia, that affect the lives of non-heterosexual people. Examples are same-sex kiss-ins in heterosexual environments, the "outing" of privileged public figures who were known or widely rumoured to be gay or lesbian and the disruption of religious services to protest heterosexist church practices. The significance of the term "nation" in the name is more rhetorical and symbolic than substantive ... Queer Nation has been much less active in the late 1990s than it was in the early 1990s' (Knopp, in McDowell and Sharp, 1999: 225).

Over the course of the 1980s and 1990s, there has been a proliferation of academic writings on dissident sexualities, spanning disciplines across the humanities and social sciences (see Abelove *et al.*, 1993, for a wide-ranging collection of interdisciplinary work in lesbian and gay studies). Much of this work has been associated with 'queer theory', which challenges the categories of 'gay', 'lesbian', 'homosexual' and 'heterosexual', by queering *all* sexualities, destabilising gendered and sexual binaries, and addressing questions of identity in much more complex and shifting terms. A great deal of research on dissident sexualities has focused on the intersections of power, identity and representation. Geographers have concentrated on the ways in which such intersections are also spatialised, focusing on sexualities over space and the critical connections between sexual spaces and sexual identities.

Sexing geography

The links between sexuality, power and knowledge are inherently spatial. The regulation of sexuality through legislation such as the Wolfenden Act in Britain continues to be couched in spatially specific terms, decriminalising certain forms of gay male sex only in spaces that are defined as 'private'. Such legislation continues to define same-sex relations between men as different from the

acceptable social norm and as something to be k̶
mind from the public dominance of heterosexuali͡
uality are also inherently spatial. The early biolog
schfield defined a 'third' or 'intermediate' sex loc
desire of heterosexuality; the psychoanalytic the
predicated on distancing, proximity and the g
sexuality in discursive terms was historically
particular contexts. Dissident sexual politic
While Germany was the main centre for lesł
twentieth century, the same could be said ⸱
twentieth century. From 1969, the gay liber⸝
and gay identities as public and visible souⸯ
and heterosexism in all spheres of life.

Although the links between sexuality, power and knowleag̶
geographical, it has only been in recent years that geographers have begu͡
study sexuality in any great detail. Important antecedents exist, as shown by
the writings of Sir Richard Burton (1821–1890), who was a well-known imperial travel writer and sexologist as well as a fellow and Gold Medallist of the
Royal Geographical Society (see Phillips, 1999, for a discussion of Burton's
mapping of a 'Sotadic Zone'). But despite Burton's geographical sexology and
vivid sexual geographies, it has taken about a century for the study of sexuality
to be taken seriously within the discipline of geography. In the more recent
past, although geographers began to examine the spatiality of class and gender
relations in response to politics of dissent in the 1960s and 1970s, it has taken
longer for geographers to reflect the importance of dissident *sexual* politics.
As Bell and Valentine write,

> The women's, gay and civil rights movements emerged in North America and
> Europe in the 1960s and 1970s on a wave of social and political upheaval. But
> despite a growing awareness amongst geographers in the following decade of
> the need to study the role of class, gender and ethnicity in shaping social,
> cultural and economic geographies, sexualities were largely left off the
> geographical map. (Bell and Valentine, 1995: 4)

By the late 1970s and early 1980s, some geographical work had begun to
address sexuality, usually by locating and mapping gay residential areas in
American cities (see, for example, Weightman, 1981; Castells, 1983; Castells
and Murphy, 1982). But, in McNee's memorable phrase, geographers were
more likely to be 'squeamish' about the sexuality of places and spaces:

> Every major metropolitan area contains spaces about which geographers tend
> to be squeamish. Consequently, such places remain unstudied or are given
> only cursory attention, at most. An approach to the study of social life less
> restricted by traditional social constraints could yield significantly scholarly
> advance. (McNee, 1984: 16)

By the late 1980s and early 1990s, more geographers were writing about
sexuality, often moving away from early attempts to locate and map gay

as to address more complex questions concerning spatiality, power. Work by David Bell, Jon Binnie, Lawrence Knopp and Gill has been particularly important in stimulating further research about geographies (see, for example, essays in the following edited volumes: e, 1994; Bell and Valentine, 1995; and Ingram *et al.*, 1997). With a few ortant exceptions (including Kramer, 1995; Valentine, 1997; and Phillips *et* ., 2000), geographical research on sexuality still tends to focus on urban space. In an attempt to make dissident sexualities visible and to resist their oppression, geographies of sexuality have largely explored the lives of gay men and lesbians. Geographers have also, but to a far lesser extent, studied bisexuality (Bell, 1994; Hemmings, 1995, 1997; also see Rose, 1996) and heterosexuality (McDowell and Court, 1994; McDowell, 1995).

Work on sexuality and space poses important challenges to geographical discourse and to geography as a discipline. Such work exposes, resists, and seeks to overturn the homophobia and heterosexism that exists within as well as beyond the academy. David Bell points to examples of censorship and discrimination that seek to constrain work on sex, sexuality and geography and to maintain an unspoken assumption of heterosexuality within geography: 'having our articles pulled from library collections, gaining negative press coverage when we get "public money" to do our work, having secretaries refuse to type up papers, not to mention all the whispering and all the silences from colleagues' (Bell, 1995: 127). Despite recent calls to situate and to embody geographical knowledge, the personal and professional costs of doing so in terms of sexuality have often been high. As Louise Johnson wrote in 1994, 'I've agonised for years about the consequences – professional and otherwise – of "coming out" in print, declaring my own sexuality and building a feminist geography upon my lesbianism. And basically I've seen the risks as too great, the stakes as too high in a homophobic culture and discipline' (Johnson, 1994: 110). Such costs and risks have been graphically detailed by Gill Valentine in her personal geography of the harassment she has suffered in the form of hate mail, silent telephone calls, threatening messages and being 'outed' to her parents. Much of this harassment has been homophobic, but Valentine views it *not* 'as an expression of a heterosexual discipline trying to purify itself, which is how I think the author intended me to read it' but rather 'as the work of a critical geographer appropriating discourses of homophobia in order to try to force me out of the discipline for personal motives' (Valentine, 1998: 316).

Many feminist geographies and geographies of sexuality stress the importance of embodying knowledge and thereby destabilising a split between mind and body, which has served to privilege the abstractions of the former over and above the materiality of the latter (see Chapter 3 for more on embodied geographies). Jon Binnie makes an important distinction between the ways in which geographers have begun to write about sexuality and the ways in which they write about sex. He argues that largely abstract geographies of sexuality have been tolerated to a greater extent than more materially embodied geographies of sex. As he writes:

> One wonders why the pleasures of the body remain off-limits while geographers of sexuality have explored the meaning of identity and community. I would posit that the tolerance afforded queer geographers tends to evaporate when confronted with the materiality of queer sex itself. Heterosexuals seem to cope with queer theory in its most abstract, intellectualised, disembodied form, but tend to run scared when confronted with the materiality of lesbian, bisexual, and gay lives, experiences, and embodiments. Despite the recent blossoming in production of literature on sexuality there is still a degree of *embarrassment* when sex is mentioned within an academic audience or text. This discomfort is exaggerated when confronted with sexual practices such as sadomasochistic sex. (Binnie, 1997: 227; also see Binnie, 1994)

Binnie calls for a 'queer epistemology' in geography, which is concerned with *how* as well as *what* geographers study. Binnie likens current work on sexuality and space to the early stages of feminist geography, *adding* lesbians and gay men to geographical analysis in similar ways to which women were added to geographical analysis in the early 1980s: 'We are still at the stage where the primary concern must be to establish lesbians and gay men as subjects and objects of geographical research ... However, it is my contention that there is a fair amount of *stirring* which still needs to be done' (Binnie, 1997: 232). For Binnie, a queer epistemology would challenge the heterosexist assumptions that persist in geographical research and teaching today. A queer epistemology would disrupt the distancing strategies associated not only with positivist spatial science and its calls for objective and disembodied analysis, but also associated with the tolerance of 'difference' and attempts to recover experiences on the 'margins' in recent social and cultural geographies (see Brown, 1995, 1997, for a critique of the distancing and erasures of medical geography and spatial science in the study of AIDS; see Bell, 1994, for a critical discussion of sexual 'margins'; and see Porteous, 1986, who suggests that geographers could usefully counter their remote and distanced analysis with some 'intimate sensing'). For Binnie, a queer epistemology would reflect the value, importance and style of camp: 'camp can productively work to undermine accepted values and truths, specifically the heterosexual definition of space. It can facilitate the creation of sexual dissident visibility in space in oppressive locations and circumstances ... Moreover in overcoming social and cultural invisibility, camp is the social practice by which sexual dissidents produce queer space through overcoming distance' (Binnie, 1997: 229–230; also see Bell, 1995; Travers, 1993).

Geographies of desire span the diverse and dynamic interfaces of sexuality and space whereby 'sexuality is implicated in the spatial construction of society and, simultaneously ... space and place are implicated in the constitution of sexual practices and sexual identity' (Knopp, 1992: 652). Sexualities, sexual practices and sexual identities are all embodied and inherently spatial from local to transnational scales. A great deal of research has explored *sexualities over space*, as shown by studies of sexuality and imperial power and the ways in which imperial representations of places and people as 'other' and exotic were often eroticised as well as racialised (see, for example, Ballhatchet, 1980;

Hyam, 1990; McClintock, 1995; and Chapter 5 of this volume) and *comparative studies* of sexualities in different places (including Binnie, 1995; Bouthillette, 1997; Munt, 1995). Important work has also explored the critical connections between *sexual spaces and identities* in everyday life, contextualising these connections in historically and geographically specific terms. As Elspeth Probyn puts it, 'Space is a pressing matter and it matters which bodies, where and how, press up against it. Most important of all is *who these bodies are with*: in what historical and actual spatial configuration they find and define themselves' (Probyn, 1995: 81; emphasis added).

Sexual spaces and identities

According to Bell and Valentine, 'In the early 1990s, geographical work, particularly in the UK, has turned away from an obsession with defining and locating gay residential and institutional communities towards a concern with identity politics' (Bell and Valentine, 1995: 8). Rather than view space as an unproblematic backdrop for sexual identities, such work has begun to interrogate the *production* of spaces as well as sexualities, and the important connections between them. Moving away from an attempt merely to map gay areas, such work examines the ways in which certain sexual identities are performed and made visible in certain places, while others are restricted and kept hidden from view. While early work tended to apply theories of racial segregation to the urban geographies of sexual dissidence, rarely interviewing people living in such places, more recent work has been largely qualitative, with archival and ethnographic research, interviews and focus groups helping to represent the rich diversity of everyday life in particular places. The work of Lawrence Knopp in the United States and Gill Valentine in Britain has been particularly important in analysing gay and lesbian spaces and identities in contemporary cities.

Lawrence Knopp has studied the intersections of capitalism, sexuality and urban space. In his study of a gay enclave in New Orleans called Marigny, Knopp explored the roles of gay businesses in attracting capital investment to the area, which resulted in gentrification and the preservation of a historic area (Knopp, 1990). He argues that such capital investment did more than political organisation to strengthen gay activism in the area. A visibly gay enclave developed as a result of middle-class entrepreneurship, and came to be marked by bookshops, cafés, and an annual Gay Fest to mark the Stonewall riot (also see Binnie, 1995, on the commercial significance of the 'pink pound' in Soho in London and in Amsterdam, and see Quilley, 1997, for an account of the entrepreneurial development of Manchester's Gay Village). Knopp has also written about the intersections of gender, sexuality and class in capitalist society, arguing that all are fundamental to identity formation and that all are dependent on unequal power relations (Knopp, 1992). Such social relations are built into the spatial ordering of society with, for example, discourses of family and domestic life shaping homes, suburbs and workplaces and infusing such spaces with heterosexual meanings that come to be naturalised and assumed. In this heterosexist context, many lesbians and gay men live in fear of their sexuality being discovered, and come to discipline themselves and their behaviour in space. By creating certain gay spaces, whether on the scale of a pub or club, a street, or a neighbourhood (such as the Castro district in San Francisco, the West End in Vancouver, Manchester's Gay Village, and Old Compton Street in London's Soho; see Figure 4.5), gay men in particular can live in greater freedom and can assert their sexuality in more publicly visible ways (also see Grube, 1997, and Hertz *et al.*, 1997, for accounts of the production of gay urban spaces in Toronto and New York). Knopp has continued to examine the sexual coding of urban space, arguing that cities come to be inscribed as both liberating and threatening to the sexual order of society (Knopp, 1995a). He points to a contradiction between cities as sites of sexual transgression and as sites of sexual control. The complexity and anonymity of modern cities may provide greater opportunities for sexual dissidence, and certain streets and districts may come to be identified as gay areas. But rather than view these areas purely as places of sexual dissidence and transgression, Knopp suggests that they also lead many people to believe that a gay subculture is located, bounded and contained, displaced far away from the home and the suburbs that remain resolutely heterosexual. In this way, Knopp argues that the visibility of gay identities in certain parts of some cities may be liberating in many ways, but, by displacing and bounding dissident sexualities, this may also appear to limit any perceived threat to the heterosexist norm of society (Knopp, 1995a).

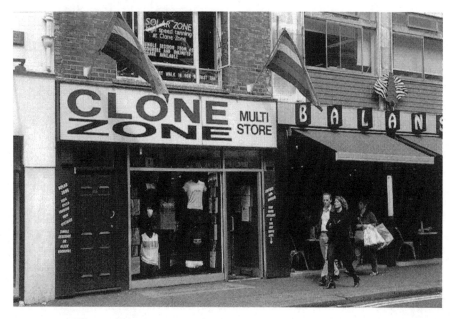

Figure 4.5 Old Compton Street, London

visible / invisible gaze

In contrast to Knopp's focus on the urban lives of gay men, Gill Valentine has examined the less visible networks linking lesbians over space. While important research has examined the production of certain lesbian spaces (see Nestle, 1997; Wolfe, 1997; and Retter, 1997, on lesbian bars and beaches and on lesbian spaces in Los Angeles), it is usually assumed that lesbian sexuality is less visible than the sexuality of many of their gay male counterparts. Valentine has explored the ways in which lesbians are often forced to conceal their sexual identity in certain places and spaces (see Figure 4.6), pretending to be heterosexual or as asexual as possible to avoid discrimination and prejudice (Valentine, 1993a, 1993b, 1993c). Networks between lesbians provide important channels of support, but may be easier to foster in some places than others, particularly in larger towns and cities than in rural areas, villages and small towns. Valentine shows how many lesbians behave differently in different places. So, for example, many women whom she interviewed had not come out as lesbians at work, and often socialised away from where they lived and worked to avoid being seen with a partner. Many lesbians were under great pressure to live separate and clearly demarcated lives at home and at work, disciplining their own behaviour over space and time in more complex ways than many of their heterosexual colleagues who could bring partners to social events and socialise with other heterosexual couples. Valentine has also written

Figure 4.6 The lesbian pink cross code

about her own experiences of occupying a split and paradoxical position: 'while I have been held up as a "lesbian–geographer" who is assumed to be "out" both publicly and privately, I have actually been performing a very different identity to my family, creating a very precarious "public"/"private", "work"/"home" splintered existence' (Valentine, 1998: 307).

Ideas about *performativity* have been particularly important in representing sexual spaces and identities, often inspired by the work of Judith Butler in theorising sex as gendered, gender as sexed, and the material embodiments of identities (Butler, 1990, 1993, 1994). Rather than assume an originary and essential sexual identity – the *essence* of a person's sexuality – ideas about performativity reveal the mobility, flux and dynamism of different identities in different contexts. Moreover, rather than assume an originary and essential *spatiality* – a stage on which different identities are played out – different spatialities and identities are seen to be mutually constituted and performed. Bell *et al.* have focused on the hypermasculinity of gay skinheads and the hyperfemininity of 'lipstick lesbians' to explore the ways in which heterosexual gender roles can be parodied and disrupted (Bell *et al.*, 1994; also see the critical commentaries on this article by Lisa Walker, Elspeth Probyn, Lawrence Knopp and Andrew Kirby, published together in 1995). Bell *et al.* argue that the performativity of sexual identities should be analysed in spatial terms. In other words, dissident sexual identities disrupt heterosexuality in ways that are inseparably bound up with the disruption of heterosexist space, subverting the norms not only of heterosexual identities but also of heterosexual spaces. For Bell *et al.*, such examples of sexual performativity are empowering as they destabilise heterosexist assumptions and help to create new, queer spaces. As they write, 'the last thing any straight person expects skinheads to do is to hold hands in public, or to gently kiss. When this happens, people notice. By behaving in this way the gay skinhead can disrupt or destabilise not only a masculine identity but heterosexual space' (Bell *et al.*, 1994: 36). Bell *et al.* interpret the hyperfemininity of lipstick lesbians as a parody of heterosexual representations of femininity by re-engaging with femininity in new, queer terms: '"Lipstick style" … has the potential to make heterosexual women question how their own appearance is read, to challenge how they see other women and hence to undermine the production of heterosexual space. Similarly the disruption of space in this way means that heterosexual men may be unable to distinguish the object of their desire – heterosexual women – from the object of their derision – the demonised "dungaree wearing lesbian"' (Bell *et al.*, 1994: 42).

But the political potential of such parodic performances has been questioned. It has been argued that Bell *et al.* represent identities in a caricatured way, with too great an emphasis placed on dress and appearance and an artificial fixity imposed on identities that are in fact complex and fluid. There are also clear disparities between the figure of the gay skinhead and the lipstick lesbian, with the former appearing to be far more successful in disrupting and reclaiming heterosexist space while the latter remains more solitary and hidden from view (Probyn, 1995). It is also important to examine the other power relations involved in parodying heterosexual identities and the ways in which they interact with gendered and sexual power relations. So, for example, it is vital to consider the ways in which such parodies of heterosexuality are racialised in particular ways and how, to many people, the style of a gay skinhead is a symbol of oppression, evoking white supremacist and fascist politics (Walker, 1995).

While Bell *et al.* have explored the ways in which dissident sexualities can parody heterosexuality, Linda McDowell has examined the ways in which *heterosexual* identities are themselves performed in embodied ways. In her study of merchant banks in the City of London, McDowell writes that work as well as social relations between men and women are constructed in heterosexist ways, based on ideas about sexual attraction and desire for the opposite sex (McDowell, 1995; also see McDowell and Court, 1994). This heterosexism not only serves to marginalise gay men and lesbians, but also serves to fix gender difference in heterosexist and masculinist ways. McDowell argues that women are assumed to be sexually active and that they are imagined more as sexual partners, wives and mothers than as colleagues who are equal to men in the workplace. For example, two women who work in merchant banks represent the unequal status of men and women in a heterosexist environment:

> If you see two or three of the girls in the department standing together having a chat, somebody will always comment 'Oh, mothers' meeting.' If you see four blokes together nobody even bats an eyelid. (Quoted in McDowell, 1995: 83)

> For most of the time they [male colleagues] treat me as an honorary male, and that's fine, I much prefer it but it also means that I see the way they look on women. If I go out for a drink with them then they will comment on anything that walks past in a short skirt ... But I guess I'd rather be the honorary male and then not have all the comments than be on the other side. (Quoted in McDowell, 1995: 83)

McDowell argues that workplaces such as merchant banks are pervaded by a heterosexist masculinity, whereby men can represent themselves as disembodied and rational while women are often represented as embodied objects of desire rather than as equals. Such heterosexist masculine norms not only marginalise women, but also exclude non-heterosexual men. It is important to study the sexual assumptions and norms that exist in workplaces to expose the ways in which heterosexuality comes to be naturalised beyond as well as within the nuclear family, home and suburb. Geographies of sexuality should be concerned not only with making dissident sexualities visible and resisting their marginalisation, but also with destabilising the assumption – and the assumed *transparency* – of heterosexuality by asking how and why opposite-sex desires have come to be naturalised and by examining the sexed nature and the pervasive power of heterosexuality throughout society.

Conclusions

Sexuality is, and always has been, an important part of human life, but has only recently come to be studied by geographers. Sexuality and space are closely intertwined, as shown by spatial explorations of sexuality in biological, psychoanalytic and discursive terms; the geographies of gay liberation, based in Germany in the early twentieth century and in the United States from the mid-twentieth century; and geographical work on the critical connections between sexual spaces and identities. Other geographical work has stressed the disciplinary implications of studying sexuality, challenging the heterosexism that exists both within and beyond the academy. The chapter began by exploring the dominance of heterosexuality, its implications for fixing gender as well as sexual difference between men and women, its power within but also beyond the nuclear family and the home, and the ways in which other sexualities are often represented as marginal and deviant to the heterosexual norm. Then the chapter examined the links between sexuality, power and knowledge, introducing the early sexological work of Ulrichs, Hirschfield and Ellis, the pscyhoanalytic theories of Freud and Lacan, and the discourse analysis of Foucault. The work of Ulrichs and Hirschfield was closely tied to sexual politics in Germany in the late nineteenth and early twentieth century. By the mid-twentieth century, the United States had become most closely associated with gay liberation campaigns. The Stonewall riot of 1969 inspired gay liberation movements in many other advanced capitalist countries. Rather than aim for integration into heterosexual society, these movements were characterised by the visible and assertive affirmation of gay identities in terms of gay pride and 'coming out'. While the work of many geographers was inspired by class and gender politics from the early 1970s onwards, it took longer for sexuality to be considered as an important and serious subject of geographical teaching and research. Early geographical studies of sexuality appeared in the 1980s and such work has become increasingly important over the course of the 1990s and will continue to challenge the heterosexism that exists in geographical discourse and geography as a discipline. Geographies of sexuality tend to focus to a great extent on the lives of lesbians and gay men, reflecting the importance of dissident sexual politics in striving for equality and sexual freedom. Geographies of sexuality have also begun to study heterosexual relations between men and women, resisting the assumed transparency and dominance of heterosexuality as a social norm. Although it remains politically and analytically imperative to make dissident sexualities visible and to resist their marginalisation, it is also important to examine the basis and effectiveness of heterosexual dominance and to study the sexed nature of heterosexuality.

References

ABELOVE, H., BARALE, M. and HALPERIN, D. (eds) (1993) *The Lesbian and Gay Studies Reader.* New York: Routledge.

BALLHATCHET, K. (1980) *Race, Sex and Class Under the Raj.* London: Weidenfeld and Nicolson.

BELL, D. (1994) Bi-sexuality: a place on the margins. In S. Whittle (ed.), *The Margins of the City: gay men's urban lives.* Aldershot: Arena, 129–141.

BELL, D. (1995) [*screw*]ing geography (censor's version). *Environment and Planning D: Society and Space*, **13**: 127–131.

BELL, D., BINNIE, J., CREAM, J. and VALENTINE, G. (1994) All hyped up and no place to go. *Gender, Place and Culture*, **1**: 31–48.

BELL, D. and VALENTINE, G. (eds) (1995) *Mapping Desire: geographies of sexualities*. London: Routledge.

BINNIE, J. (1994) The twilight world of the sadomasochist. In S. Whittle (ed.), *The Margins of the City: gay men's urban lives*. Aldershot: Arena, 157–169.

BINNIE, J. (1995) Trading places: consumption, sexuality and the production of queer space. In D. Bell and G. Valentine (eds), *Mapping Desire: geographies of sexualities*. London: Routledge, 182–199.

BINNIE, J. (1997) Coming out of geography: towards a queer epistemology? *Environment and Planning D: Society and Space*, **15**: 223–237.

BLASIUS, M. and PHELAN, S. (eds) (1997) *We Are Everywhere: a historical sourcebook of gay and lesbian politics*. New York: Routledge.

BOUTHILLETTE, A.-M. (1997) Queer and gendered housing: a tale of two neighbourhoods in Vancouver. In G.B. Ingram, A.-M. Bouthillette and Y. Retter (eds), *Queers in Space: communities/public places/sites of resistance*. Seattle, WA: Bay Press, 213–232.

BROWN, M. (1995) Ironies of distance: an ongoing critique of the geographies of AIDS. *Environment and Planning D: Society and Space*, **13**: 159–183.

BROWN, M. (1997) *Replacing Citizenship: AIDS activism and radical democracy*. New York: Guilford.

BROWN, M. (1999) Travelling through the closet. In J. Duncan and D. Gregory (eds), *Writes of Passage: reading travel writing*. London: Routledge, 185–199.

BUTLER, J. (1990) *Gender Trouble: feminism and the subversion of identity*. New York: Routledge.

BUTLER, J. (1993) *Bodies that Matter: on the discursive limits of 'sex'*. New York: Routledge.

BUTLER, J. (1994) Gender as performance: an interview with Judith Butler. *Radical Philosophy*, **67**: 32–39.

CASTELLS, M. (1983) *The City and the Grassroots*. Berkeley, CA: University of California Press.

CASTELLS, M. and MURPHY, K. (1982) Cultural identity and urban structure: the spatial organisation of San Francisco's gay community. In N.I. Fanstein and S.S. Fanstein (eds), *Urban Policy under Capitalism*. Beverly Hills, CA: Sage.

DE LAURETIS, T. (1994) *The Practice of Love: lesbian sexuality and perverse desire*. Bloomington: Indiana University Press.

D'EMILIO, J. (1989) Gay politics and community in San Francisco since World War II. In M. Duberman, M. Vicinus and G. Chauncey (eds), *Hidden from History: reclaiming the gay and lesbian past*. New York: Meridian.

D'EMILIO, J. and FREEDMAN, E. (1988) *Intimate Matters: a history of sexuality in America*. New York: Harper and Row.

DIAMOND, I. and QUINBY, L. (eds) (1988) *Feminism and Foucault.* Boston, MA: Northeastern University Press.

DUBERMAN, M. (1994) *Stonewall.* New York: Plume.

DUBERMAN, M., VICINUS, M. and CHAUNCEY, G. (eds) (1989) *Hidden from History: reclaiming the gay and lesbian past.* New York: Meridian.

FOUCAULT, M. (1973) *The Birth of the Clinic.* London: Tavistock.

FOUCAULT, M. (1990) [1976] *The History of Sexuality,* vol. 1. New York: Vintage.

GARBER, E. (1989) A spectacle in color: the lesbian and gay subculture of Jazz Age Harlem. In M. Duberman, M. Vicinus and G. Chauncey (eds), *Hidden from History: reclaiming the gay and lesbian past.* New York: Meridian.

GRUBE, J. (1997) 'No more shit': the struggle for democratic gay space in Toronto. In G.B. Ingram, A.-M. Bouthillette and Y. Retter (eds), *Queers in Space: communities/public places/sites of resistance.* Seattle, WA: Bay Press, 127-146.

HAEBERLE, E. (1989) Swastika, pink triangle, and yellow star: the destruction of sexology and the persecution of homosexuals in Nazi Germany. In M. Duberman, M. Vicinus and G. Chauncey (eds), *Hidden from History: reclaiming the gay and lesbian past.* New York: Meridian.

HASTE, C. (1994) *Rules of Desire: sex in Britain, World War I to the present.* London: Pimlico.

HEALEY, E. and MASON, A. (eds) (1994) *Stonewall 25: the making of the lesbian and gay community in Britain.* London: Virago.

HEMMINGS, C. (1995) Locating bisexual identities: discourses of bisexuality and contemporary feminist theory. In D. Bell and G. Valentine (eds), *Mapping Desire: geographies of sexualities.* London: Routledge, 41–55.

HEMMINGS, C. (1997) From landmarks to spaces: mapping the territory of a bisexual genealogy. In G.B. Ingram, A.-M. Bouthillette and Y. Retter (eds), *Queers in Space: communities/public places/sites of resistance.* Seattle, WA: Bay Press, 147–162.

HERTZ, B.-S., EISENBERG, E. and KNAUER, L.M. (1997) Queer spaces in New York City: places of struggle/places of strength. In G.B. Ingram, A.-M. Bouthillette and Y. Retter (eds), *Queers in Space: communities/public places/sites of resistance.* Seattle, WA: Bay Press, 357–370.

HYAM, R. (1990) *Empire and Sexuality: the British experience.* Manchester: Manchester University Press.

INGRAM, G.B., BOUTHILLETTE, A.-M. and RETTER, Y. (eds) (1997) *Queers in Space: communities/public places/sites of resistance.* Seattle, WA: Bay Press.

JOHNSON, L. (1994) What future for feminist geography? *Gender, Place and Culture,* 1: 103–114.

JOHNSTON, R.J., GREGORY, D. and SMITH, D.L. (eds) (1994) *The Dictionary of Human Geography.* Oxford: Blackwell.

KIRBY, A. (1995) Straight talk on the PomoHomo question. *Gender, Place and Culture,* 2: 89–96.

KNOPP, L. (1990) Some theoretical implications of gay involvement in an urban land market. *Political Geography Quarterly,* **9**: 337–352.

KNOPP, L. (1992) Sexuality and the spatial dynamics of late capitalism. *Environment and Planning D: Society and Space,* **10**: 651–669.

KNOPP, L. (1995a) Sexuality and urban space: a framework for analysis. In D. Bell and G. Valentine (eds), *Mapping Desire: geographies of sexualities.* London: Routledge, 149–161.

KNOPP, L. (1995b) If you're going to get all hyped up, you'd better go *somewhere. Gender, Place and Culture,* **2**: 85–88.

KRAMER, J.L. (1995) Bachelor farmers and spinsters: gay and lesbian identities and communities in rural North Dakota. In D. Bell and G. Valentine (eds), *Mapping Desire: geographies of sexualities.* London: Routledge, 200–213.

LORDE, A. (1984) *Sister Outsider: essays and speeches.* Trumansburg, NY: Crossing Press.

McCLINTOCK, A. (1995) *Imperial Leather: race, gender and sexuality in the colonial contest.* New York: Routledge.

McDOWELL, L. (1995) Body work: heterosexual gender performances in city workplaces. In D. Bell and G. Valentine (eds), *Mapping Desire: geographies of sexualities.* London: Routledge, 75–95.

McDOWELL, L. and COURT, G. (1994) Missing subjects: gender, power and sexuality in merchant banking. *Economic Geography,* **70**: 229–251.

McDOWELL, L. and SHARP, J. (eds) (1999) *A Feminist Glossary of Human Geography.* London: Arnold.

McNEE, B. (1984) If you are squeamish .... *East Lakes Geographer,* **19**: 16–27.

MILLIGAN, D. (1993) *Sex Life: a critical commentary on the history of sexuality.* London: Pluto.

MUNT, S. (1995) The lesbian *flâneur.* In D. Bell and G. Valentine (eds), *Mapping Desire: geographies of sexualities.* London: Routledge, 114–125.

NAST, H. (1998) Unsexy geography. *Gender, Place and Culture,* **5**: 191–206.

NESTLE, J. (1997) Restriction and reclamation: lesbian bars and beaches of the 1950s. In G.B. Ingram, A.-M. Bouthillette and Y. Retter (eds), *Queers in Space: communities/public places/sites of resistance.* Seattle, WA: Bay Press, 61-68.

PARKER, A. *et al.* (eds) (1992) *Nationalisms and Sexualities.* London: Routledge.

PHILLIPS, R. (1999) Writing travel and mapping sexuality: Richard Burton's Sotadic Zone. In J. Duncan and D. Gregory (eds), *Writes of Passage: reading travel writing.* London: Routledge, 70–91.

PHILLIPS, R., WATTS, D. and SHUTTLETON, D. (eds) (2000) *Decentring Sexualities: politics and representations beyond the metropolis.* London: Routledge.

PLANT, R. (1988) *The Pink Triangle: the Nazi war against homosexuals.* New York: Henry Holt.

PORTEOUS, D. (1986) Intimate sensing. *Area,* **18**: 250–251.

PROBYN, E. (1995) Lesbians in space. Gender, sex and the structure of missing. *Gender, Place and Culture*, **2**: 77–84.

QUILLEY, S. (1997) Constructing Manchester's 'new urban village': gay space in the entrepreneurial city. In G.B. Ingram, A.-M. Bouthillette and Y. Retter (eds), *Queers in Space: communities/public places/sites of resistance*. Seattle, WA: Bay Press, 275–292.

RETTER, Y. (1997) Lesbian spaces in Los Angeles, 1970-90. In G.B. Ingram, A.-M. Bouthillette and Y. Retter (eds), *Queers in Space: communities/public places/sites of resistance*. Seattle, WA: Bay Press, 325–338.

RICH, A. (1984) Compulsory heterosexuality and lesbian existence. In A. Snitow, C. Stansell and S. Thompson (eds), *Desire: the politics of sexuality*. London: Virago, 212–241.

RICH, A. (1986) *Blood, Bread and Poetry: selected prose, 1979–1985*. London: Virago.

ROSE, G. (1993) *Feminism and Geography: the limits to geographical knowledge*. Cambridge: Polity.

ROSE, S. (ed.) (1996) *Bisexual Horizons: politics, histories, lives*. New York: Lawrence and Wishart.

RUSSETT, C. (1989) *Sexual Science: the Victorian construction of womanhood*. Cambridge, MA: Harvard University Press.

SAWICKI, J. (1991) *Disciplining Foucault: feminism, power and the body*. New York: Routledge.

SHOWALTER, E. (1990) *Sexual Anarchy: gender and culture at the* fin de siècle. London: Penguin.

SINFIELD, A. (1998) *Gay and After*. London: Serpent's Tail.

SONTAG, S. (1989) *AIDS and its Metaphors*. New York: Farrar Strauss.

TONG, R. (1992) *Feminist Thought: a comprehensive introduction*. London: Routledge.

TRAVERS, A. (1993) An essay on self and camp. *Theory, Culture and Society*, **10**: 127–143.

VALENTINE, G. (1993a) (Hetero)sexing space: lesbian perspectives and experiences of everyday spaces. *Environment and Planning D: Society and Space*, **11**: 395–413.

VALENTINE, G. (1993b) Negotiating and managing multiple sexual identities: lesbian time–space strategies. *Transactions of the Institute of British Geographers* (NS), **18**: 237–248.

VALENTINE, G. (1993c) Desperately seeking Susan: geographies of lesbian friendships. *Area*, **25**: 109–116.

VALENTINE, G. (1997) Making space: separatism and difference. In J.P. Jones, H. Nast and S. Roberts (eds.), *Thresholds in Feminist Geography: difference, methodology, representation*. Lanham, Maryland: Rowman and Littlefield, 65–76.

VALENTINE, G. (1998) 'Sticks and stones may break my bones': a personal geography of harassment. *Antipode*, **30**: 305–332.

WALKER, L. (1995) More than just skin-deep: fem(me)ininity and the subversion of identity. *Gender, Place and Culture*, **2**: 71–76.

WEEDON, C. (1997) *Feminist Practice and Poststructuralist Theory.* Oxford: Blackwell.

WEEKS, J. (1977) *Coming Out: homosexual politics in Britain, from the nineteenth century to the present.* London: Quartet.

WEEKS, J. (1991) *Against Nature: essays on history, sexuality and identity.* London: Rivers Oram.

WEEKS, J. (1995) *Invented Moralities: sexual values in an age of uncertainty.* Cambridge: Polity.

WEIGHTMAN, B.A. (1981) Commentary: towards a geography of the gay community. *Journal of Cultural Geography,* 1: 106–112.

WHITTLE, S. (ed.) (1994) *The Margins of the City: gay men's urban lives.* Aldershot: Arena.

WOLFE, M. (1997) Invisible women in invisible places: the production of social space in lesbian bars. In G.B. Ingram, A.-M. Bouthillette and Y. Retter (eds), *Queers in Space: communities/public places/sites of resistance.* Seattle, WA: Bay Press, 301-324.

WOLFF, C. (1986) *Magnus Hirschfield: a portrait of a pioneer in sexology.* London: Quartet.

WOOD, N. (1985) Foucault on the history of sexuality: an introduction. In V. Beechey and J. Donald (eds), *Subjectivity and Social Relations.* Milton Keynes: Open University Press, 156–174.

5

Decolonising Geography: Postcolonial Perspectives

The aims of a post-colonial geography might be defined as: the unveiling of geographical complicity in colonial dominion over space; the character of geographical representation in colonial discourse; the de-linking of local geographical enterprise from metropolitan theory and its totalizing systems of representation; and the recovery of those hidden spaces occupied, and invested with their own meaning, by the colonial underclass. (Crush, 1994: 336–337)

Postcolonialism

The term 'postcolonialism' is contested and diverse, encompassing a range of work in disciplines such as literary studies, cultural studies, anthropology and history as well as geography (see the following edited volumes, which reflect the diversity of postcolonial studies: Ashcroft *et al.*, 1995; Williams and Chrisman, 1993; Clayton and Gregory, forthcoming). Such interdisciplinary postcolonial studies have become increasingly important over the course of the 1980s and 1990s, and are likely to continue to challenge geography and other disciplines in profound and far-reaching ways in the future. As the quotation from Jonathan Crush suggests, the aims of a postcolonial geography are themselves diverse, encompassing the history as well as the present status of the discipline, the ways in which geographical imaginations have underpinned colonial power and knowledge, and the need to recover the experiences and agency of colonised peoples.

Broadly speaking, postcolonial perspectives are *anti-colonial*, exploring the impact of colonialism in the past and in the present; investigating the links between colonial forms of power and knowledge; and resisting colonialism and colonial representations of the world. Postcolonial critiques of geography as a discipline have included studies of the colonial importance and complicity of the subject as it became institutionalised and increasingly influential, particularly in Britain, France and Germany in the nineteenth and early twentieth centuries (Bell *et al.*, 1995; Driver, 1992; Godlewska and Smith, 1994; Livingstone, 1993).

Many people argue that this colonial history continues to shape geography as a discipline today, pointing to why and how certain subjects are studied and critiquing ethnocentric tendencies of western geography (Gregory, 1994). In discursive terms, geography has been shown as central to the exercise of colonial power and the production of colonial knowledge as people and places throughout the world were brought under external control and were represented in often stereotypical and derogatory ways over space (Gregory, 1995a; Said, 1978). Postcolonial perspectives highlight the *importance* of representing people and places across different cultures, traditions and contexts but also point to the *difficulties* of such endeavours. At the same time, postcolonial critiques stress the need to destabilise what might be taken for granted and assumed in our own cultures, traditions and contexts. So, for example, postcolonial studies challenge the production of knowledges that are exclusively western and ethnocentric by not only focusing on the world beyond 'the West' but also by destabilising what is understood by and taken for granted about 'the West' (Young, 1990). Postcolonial studies are concerned with the impact of colonialism on western and non-western cultures and societies and aim to 'decolonise the mind' from the ethnocentrism of dominant western culture and society. Decolonising geography is a multifaceted task, reflecting the need to reassess the history of geography; to challenge ethnocentric tendencies in geography today; to reveal the geographical underpinnings of colonial power and knowledge; to resist these geographies of colonialism and colonial knowledge; and to write postcolonial geographies that focus on people and places that have been marginalised in colonial and neo-colonial representations of the world.

This chapter will begin by introducing the contested and diverse meanings of the term 'postcolonialism'. Then the importance of imperial and colonial geographies will be discussed, tracing the differences and connections between imperialism and colonialism; different stages of imperial expansion; and the emergence of colonial discourse analysis. The influential work of Edward Said, Homi Bhabha and Gayatri Chakravorty Spivak will be introduced here. In the next section, the links between geography and imperialism will be explored, focusing on critical histories of British geography. Then the importance of decolonising geography in the present will be assessed by investigating cross-cultural representations over space today and the importance of recovering 'hidden spaces' beyond the West while, at the same time, destabilising notions of 'the West' and its centrality to the exercise of power and the production of knowledge. Finally, an example of postcolonial geographical practice will be examined with reference to the work of Action from Ireland in commemorating the Irish Famine.

Postcolonial perspectives

At its broadest, postcolonialism 'deals with the effects of colonization on cultures and societies' (Ashcroft *et al.*, 1998: 186). The effects of colonisation are studied both in the past and in the present, with many commentators arguing that present inequalities in the world have not only been shaped by a colonial legacy but also persist today because of neo-colonial power relations that continue to exploit poorer countries and regions for the material benefit of their wealthier counterparts. Recognising the persistence of neo-colonial domination, postcolonialism is 'a continuing process of resistance and reconstruction' (Ashcroft *et al.*, 1995: 3), exposing and challenging power relations that both result in and depend upon neo-colonial inequalities.

The term 'postcolonial' was first used after the Second World War as a chronological marker, referring to the 'post-independence' era that followed decolonisation. But, from the late 1970s, the term came to be significant in more than just a chronological sense, as shown by its importance in studies of Commonwealth literature and New Literatures in English (see Boehmer, 1995, for a clear and wide-ranging introduction to 'the writing of empire, and ... writing in opposition to empire': 1). Since then, the term has been widely used to refer to 'the political, linguistic and cultural experience of societies that were former European colonies' (Ashcroft *et al.*, 1998: 186), encompassing the effects of colonisation and decolonisation.

In theoretical terms, postcolonial studies have been greatly influenced by marxist and poststructuralist analyses. The differences between marxism and poststructuralism have often been caricatured as an unbridgeable gulf. But, as much postcolonial work reveals, these two diverse schools of thought are marked more by their connections and productive tensions than by a polarised hostility. Many of the earliest and most influential critiques of imperialism have been written within a marxist tradition, exploring European conquest and territorial expansion as inseparably bound up with the global extent and inequalities of late capitalism (see Chapter 2). The ideas of many poststructuralist critics developed from within a marxist tradition and share its anti-humanism and historical materialism. In the context of postcolonialism, the political-economy of marxist analyses has been important alongside the cultural and linguistic analyses of poststructuralism. Poststructuralist influences on postcolonial studies include a focus on the connections between power and knowledge, the politics of representation and questions of identity and difference (see Chapter 3 for an introduction to poststructuralism, and see Young, 1990, and Gandhi, 1998, for further explorations of the postcolonial implications of both of these schools of thought).

The 'post' of 'postcolonialism' has two meanings, referring to a temporal aftermath – *after* colonialism – and a critical aftermath – *beyond* colonialism. But these two meanings do not necessarily coincide and it is their problematic interaction that often makes 'postcolonialism' a contested term. A great deal of discussion has focused on the limits of thinking that postcolonialism refers to a period *after* colonialism. First, this temporal distinction implies a clear break with a colonial past, often obscuring the continuities in international relations that persist even after formal decolonisation might have occurred. Second, the persistence of international inequalities in a neo-colonial world throws the very possibility of decolonisation into question. Third, it can be argued that the temporal underpinnings of the term 'postcolonialism' continue to define the world purely in terms of western expansion. In other words, colonised peoples and places become the subject of study only by virtue of having been colonised. As Anne McClintock writes, the focus on temporal difference (between a colonial and a postcolonial era) comes to supersede a focus on power relations (between colonisers and colonised) and 'colonialism returns at the moment of its disappearance' (McClintock, 1995: 11).

In light of the difficulties of referring to a clearly defined postcolonial era, many writers refer to postcolonialism as being *beyond* rather than only and necessarily *after* colonialism. In this case, the 'post' in 'postcolonialism' refers more to a *critical* than to a *temporal* aftermath as postcolonial perspectives explore and resist colonial and neo-colonial power and knowledge. In this way, postcolonialism offers widely ranging critiques of colonialism which are under-pinned by anti-colonial politics. As Ania Loomba suggests, 'it is more helpful to think of postcolonialism not just as coming literally after colonialism and signifying its demise, but more flexibly as the *contestation* of colonial domina-tion and the legacies of colonialism' (Loomba, 1998: 12, emphasis added; also see Moore-Gilbert, 1997). In a similar way, Jane Jacobs writes that 'Postcolo-nialism may be better conceptualised as an historically dispersed set of forma-tions which negotiate the ideological, social and material structures of power established under colonialism' (Jacobs, 1996: 25). This chapter will stress that postcolonialism should also be understood as a *geographically* dispersed con-testation of colonial power and knowledge.

Postcolonial perspectives challenge colonial power and its legacies today. They do so in part by examining the *basis* of colonial power and the associated production of colonial knowledge. In other words, postcolonial studies address the ways in which colonial power was exercised, legitimated, resisted and over-turned over time and space. Rather than generalise about colonial power and knowledge, postcolonial critiques reveal the historical and geographical diver-sity of colonialism and the need to ground such critiques in material and spe-cific contexts. A specific location in time and space is vital because 'Every colonial encounter or "contact zone" is different, and each "post-colonial" occasion needs ... to be precisely located and analysed for its specific interplay' (Ashcroft *et al.*, 1998: 190). The effects of colonialism were not just one-way, transported from the metropolis to colonies. Rather, as Loomba writes,

> Postcolonial studies have shown that both the 'metropolis' and the 'colony' were deeply altered by the colonial process. *Both* of them are, accordingly, also restructured by decolonisation. This of course does not mean that both are postcolonial *in the same way*. Postcoloniality, like patriarchy, is articulated alongside other economic, social, cultural and historical factors, and therefore, in practice, it works quite differently in various parts of the world. (Loomba, 1998: 19)

As an important foundation for understanding postcolonial critiques, the next section will explore the spatial extent and diversity of imperialism and colonialism.

Imperial and colonial geographies

Although the terms 'imperialism' and 'colonialism' are often used interchange-ably, there are important differences between them (see Ashcroft *et al.*, 1998, for helpful definitions of these and other 'key concepts' in postcolonial studies). Imperialism refers in general terms to a system of domination over space, encompassing 'the practice, the theory, and the attitudes of a dominating metropolitan centre ruling a distant territory' (Said, 1993: 9). As one tangible manifestation of imperial power, 'colonialism' refers more specifically to 'the implanting of settlements on a distant territory' (Said, 1993: 9). Colonialism, which is 'almost always a consequence of imperialism' (Said, 1993: 9), depends on conquest, territorial expansion, and processes of colonisation whereby people, goods, and capital move from a metropolitan centre to a colony. Both colonialism and imperialism bind metropolitan centres and colonies together in an unequal system of power and dependence. But colonialism represents the direct imposition of imperial rule through settlement and political control over 'a separate group of people, who are viewed as subordinate, and their territories, which are presumed to be available for exploitation' (Jacobs, 1996: 16). Imperialism is therefore a more general term, which can refer to economic, political and cultural inequalities and dependencies whereby a country, region or group of people are subject to the rule of a separate and more powerful force] Imperial power can be exercised by nation states (such as the UK, France and Germany in the nineteenth century), companies (as shown by the transnational operations of, for example, oil companies such as Shell in Nigeria) and organisations (including the regulatory power of the World Bank and the International Monetary Fund). Within the broad parameters of imperial rule, colonialism represents the imposition of political control through conquest and territorial expansion over people and places located at a distance from the metropolitan power.

Throughout human history, imperial and colonial power has brought different parts of the world and different groups of people under external control:

> At its height in the second century AD, the Roman Empire stretched from Armenia to the Atlantic. Under Genghis Khan in the thirteenth century, the Mongols conquered the Middle East as well as China. The Aztec Empire was established when, from the fourteenth to the sixteenth centuries, one of the various ethnic groups who settled in the valley of Mexico subjugated the others. Aztecs extracted tributes in services and goods from conquered regions, as did the Inca Empire which was the largest pre-industrial state in the Americas. In the fifteenth century too, various kingdoms in southern India came under the control of the Vijaynagara Empire, and the Ottoman Empire, which began as a minor Islamic principality in what is now western Turkey, extended itself over most of Asia Minor and the Balkans. At the beginning of the eighteenth century, it still extended from the Mediterranean to the Indian Ocean, and the Chinese Empire was larger than anything Europe had seen. (Loomba, 1998: 2–3)

Such a far-reaching catalogue of territorial expansion reveals the diversity and global extent of imperial and colonial power throughout human history. But the extent and effects of modern European expansion far surpassed these previous empires. From the sixteenth to the twentieth centuries, countries such as Belgium, the UK, France, Germany, Italy, the Netherlands, Portugal and Spain were all colonial powers, with their influence extending over vast areas in Asia, Africa, the Americas and Australasia. While modern European expansion should not be viewed in isolation from earlier empires, its scale, longevity and unprecedented levels of organisation meant that it was 'a constitutively, radically different type of overseas domination from all earlier forms' (Said, 1993: 221). By the 1930s, European colonies and ex-colonies accounted for 84.6% of the globe and 'only parts of Arabia, Persia, Afghanistan, Mongolia, Tibet, China, Siam and Japan had never been under formal European government' (Loomba, 1998: xiii).

Unlike earlier empires, the colonial power of Western European countries from the sixteenth to the twentieth centuries represented the territorial expansion of capitalism, drawing distant countries into a capitalist world system of production, distribution and exchange (Corbridge, 1986). Referring to imperialism as 'the highest stage of capitalism' in 1915, Lenin argued that surplus finance-capital in western economies, where there was a lack of labour resources, would be invested in colonies where there was a lack of capital but an abundance of labour. This capital investment was seen as a necessary condition for capitalist accumulation to continue and to expand. But, at the same time, Lenin argued that imperialism represented the *final* stage of capitalism because the rivalry between different imperial powers would lead to war and, eventually, to the revolutionary overthrow of the capitalist system (Lenin, [1915] 1978; see Chapter 2 for further discussion).

| Modern European empires were capitalist empires, whereby 'the expansion of the accumulative capacities of capitalism was realised through the conquest and possession of other people's land and labour in the service of the metropolitan core'| (Jacobs, 1996: 16). But, as Loomba suggests, 'Modern colonialism did more than extract tribute, goods and wealth from the countries that it conquered – it restructured the economies of the latter, drawing them into a complex relationship with their own, so that there was a flow of human and natural resources between colonised and colonial countries' (Loomba, 1998: 3). Three main stages of modern European expansion have been identified, each associated with distinctive flows of human and natural resources between metropolitan centres and distant colonies. The first stage of European colonial expansion dates from the sixteenth century, when a crisis of feudalism, the emergence of new European states, and wars over sovereignty, led countries to search for new sources of revenue, particularly in the form of silver and gold. At this time, Spanish and Portuguese trade led to settlement in South America, the exploitation of mineral resources and the large-scale, forced conversion of indigenous peoples to Roman Catholicism. From the seventeenth century, an era of mercantile imperialism was characterised by new forms of settlement and trade which were closely linked to the development of manufacturing in Europe. The colonial expansion of Britain in North America and of Britain, France and the Netherlands in the West Indies relied on the slave trade. Between 1701 and 1810, more than six million slaves were transported from Africa to the Caribbean, Brazil and what are now the southern states of the United States (see Figure 5.1). A triangular trade linked West Africa, Europe

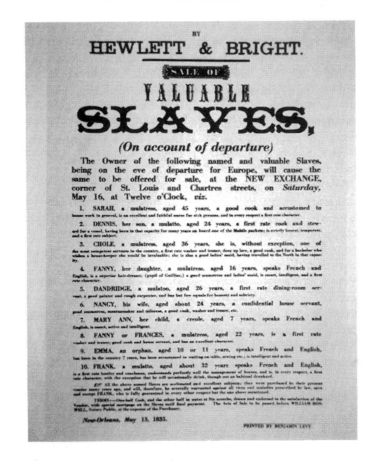

Figure 5.1 Slavery in America, New Orleans, 1835

and the 'New World' of the Americas. West African slaves were transported as labour to the Americas where they worked on plantations that produced goods such as cotton, tobacco, sugar, indigo and rum. These goods were then exported to Western Europe, where they were sold for profit or manufactured into other goods, which were then, in turn, exported to West Africa and other parts of the world. Ports in Britain such as Liverpool and Bristol played a central role in this triangular trade until the slave trade was abolished in 1807 (Gilroy, 1994; Meegan, 1995). But, despite the abolition of the slave *trade* in 1807, Britain did not outlaw slavery in its colonies until 1833. Plantation slavery persisted in the West Indies and parts of South America until the 1830s and was not outlawed in the southern United States until the victory of the North in the American Civil War. The abolition of slavery in the United States was only formally ratified in 1865 (Ashcroft *et al.*, 1998).

The third stage of European expansion lasted from the nineteenth to the early twentieth centuries, and has been called 'the age of empire' or an era of 'high imperialism' (Hobsbawm, 1989). At this time, European states played increasingly important roles in co-ordinating the expansion of capitalist imperialism. New types and unprecedented volumes of raw materials such as cotton were needed to support the rapid pace of industrialisation in Western Europe. At the same time, colonies provided extensive new markets for factory-produced commodities, and the export of such commodities was facilitated by the development of new forms of transport and communications. What Marx called 'the annihilation of space by time' accelerated as the telegraph, steamships and railways bound colonies and colonial powers into greater proximity than ever before (see Chapter 2). Among imperial powers, the British Empire was the most extensive and profitable and incorporated settler colonies as far-flung as Canada, Australia, New Zealand, Kenya and South Africa; (Figure 5.2; and see Morris, 1968, for an account of the British Empire at its height). India, which came under the direct governance of the British Crown in 1858 after three centuries of economic rule by the East India Company, was known as the imperial 'jewel in the crown'.

In many countries, the 'age of empire' ended with decolonisation and independence (see Box 5.1; Betts, 1998; Chamberlain, 1985; Darwin, 1988). In the nineteenth and early twentieth centuries, Dominion Status was achieved by settler colonies such as Canada (1867), Australia (1900), New Zealand (1907) and South Africa (1910). Elsewhere, the rising tide of nationalist and other independence movements, coupled with the political and economic consequences of the Second World War, led to the large-scale dismantling of Western European empires from the late 1940s. British India was partitioned into India and Pakistan, which became independent nations in 1947, as discussed in Box 5.2. As Chamberlain writes,

> [British India] was the largest single country decolonized, as well as the first important example of decolonization after 1945. ... [I]t seems beyond dispute that India was the great exemplar to which colonial nationalists in other countries looked and that the relinquishment of India in 1947 set the British empire (by far the largest of the European colonial empires) inexorably upon the path to dissolution. (Chamberlain, 1985: 1–2)

Many other colonies did not become autonomous and independent until the 1960s and, most recently, Hong Kong was handed over from the United Kingdom to China at the end of a 99-year lease in 1997.

Box 5.1 Decolonisation

As Betts writes, 'decolonization is most easily appreciated and measured as a series of political acts, occasionally peaceful, often confrontational, and frequently militant, by which territories and countries dominated by Europeans gained their independence' (Betts, 1998: 98). Although the term 'decolonisation' was first used in the 1930s, it came into more general use only in the 1950s and 1960s (Chamberlain, 1985). Many people would argue that the term is problematic, implying that the initiatives for decolonisation were taken by the metropolitan, ruling powers rather than by colonised peoples themselves. As Chamberlain shows, decolonisation varied over time and space and usually

Box 5.1 *continued*

involved 'the policies of the colonial powers and the ideas and initiatives which came from the colonized' (Chamberlain, 1985: 1). Figure 5.1.1 maps the political decolonisation of the main colonies and colonial territories over the course of the twentieth century. Several of these places changed their names after independence and their new names appear in parentheses. As the map shows, there were three main periods of political decolonisation: the late 1940s in Southeast Asia, the 1950s in North Africa, and the 1960s in Sub-Saharan Africa.

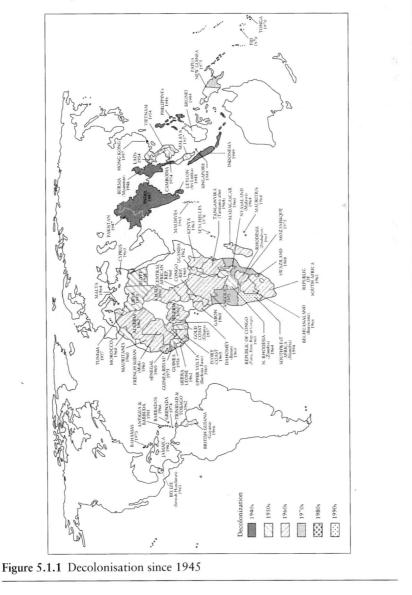

Figure 5.1.1 Decolonisation since 1945

Figure 5.2 The extent of the British Empire, 1902 (G. Amato in the *Illustrated London News* special issue for the coronation of Edward VII)

Box 5.2 British India: independence and partition in 1947

The Indian subcontinent achieved independence from British rule at midnight on 14/15 August, 1947. Independence was achieved at the expense of a united India. The partition of the subcontinent into a largely Hindu India and a largely Muslim East and West Pakistan (Pakistan and Bangladesh since 1971) led to the migration of up to 14 million people, communal conflict, and the deaths of about one million people (see Butalia, 1998, and Menon and Bhasin, 1998, for more on partition). According to Patrick French,

> few moments in modern history have had a more lasting impact on so many people. As the writer and politician André Malraux suggested, Britain's decision to get out of India was 'the most significant fact of the century'. It removed three-quarters of King George VI's subjects overnight, reduced Britain to a 'third rate power', and proved that the practice of European imperialism was no longer sustainable.
>
> The nature of the political settlement in 1947 had a calamitous impact on the subcontinent, leading to the reciprocal genocide and displacement of millions of Hindus, Muslims and Sikhs, three Indo-Pakistan wars, the blood-drenched creation of Bangladesh, and the long-term limitation of the region's global influence. Although more than a fifth of the world's population presently lives in the territory of Britain's former Indian Empire, continued internal conflict has left South Asia with little cohesiveness and minimal international clout. Nearly half of all Pakistani government expenditure still goes on the cold war with India, focused on the running sore of Kashmir. (French, 1998: xxii–xxiii)

One of the most important figures in the struggle for independence was Mahatma Gandhi, leader of the Indian National Congress (see Figure 5.2.1). Gandhi worked as a lawyer in South Africa for 22 years, where he fought racist legislation and developed his faith in the power of passive resistance or 'satyagraha'. He returned to India in 1914, aged 45, where he supported the British throughout the First World War. The massacre at Jallianwala Bagh in Amritsar in 1919 proved a turning point in the struggle for self-rule and in Gandhi's leadership. General Dyer ordered his soldiers to open fire on a crowd that had gathered in this public courtyard, killing at least 300 Indians and injuring a further 1500. After a public inquiry, Dyer was dismissed from the army, but he was hailed as an imperial hero by many people in Britain. In India, the Indian National Congress and the Muslim League escalated their campaigns for independence or 'swaraj'. Gandhi told his followers that 'Co-operation in any shape or form with this satanic government is sinful' (James, 1997: 154).

Although Gandhi was extremely influential in his non-violent campaigns and his calls for communal harmony, the struggle for independence was often violent and bloody and included bombings, assassinations, riots and the mass imprisonment of many thousands of people. Gandhi was assassinated by a Hindu extremist in January 1948, who blamed him for the partition of India. As French writes, 'For all his failings and contradictions, he remains the central figure in India's journey to independence and division, and an iconic leader in twentieth-century world history' (French, 1998: 361). In her biography of Gandhi, Judith Brown writes that he may not have found

Box 5.2 *continued*

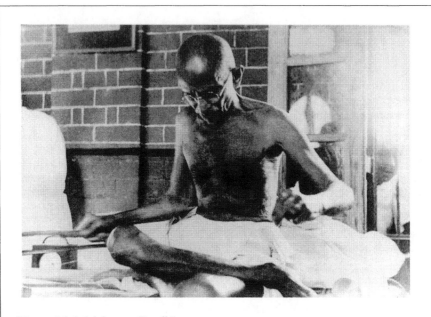

Figure 5.2.1 Mahatma Gandhi

lasting and real solutions to many of the problems he encountered. Possibly he did not even see the implications of some of them. ... [Yet he] asked many of the profoundest questions that face humankind as it struggles to live in community [and it is this] which marks his true stature and which makes his struggles and glimpses of truth of enduring significance. As a man of his time who asked the deepest questions, even though he could not answer them, he became a man for all times and all places. (Brown, quoted in French, 1998: 362)

Although the formal structures of colonial rule might have been overturned, the legacies of colonial rule remain intact in many spheres of life both in metropolitan centres and in ex-colonies. The political, administrative, legal, educational and religious systems in many ex-colonies continue to reflect past European colonial influence. As Box 5.3 and Figure 5.3 show, the Commonwealth is an international association that links many countries that had been part of the British Empire. In economic terms, colonial rule often led to regional specialisation so that regions, or even whole countries, became focused on producing a specific raw material or food crop for export, as shown by the commodity dependence on bananas in Central America, tea in Sri Lanka and rubber in Malaya (Enloe, 1990). But this dependence on export and global exchange has made regions and countries vulnerable to crop failures, price fluctuations and changes in international demand, continuing to bind ex-colonies to the crises of

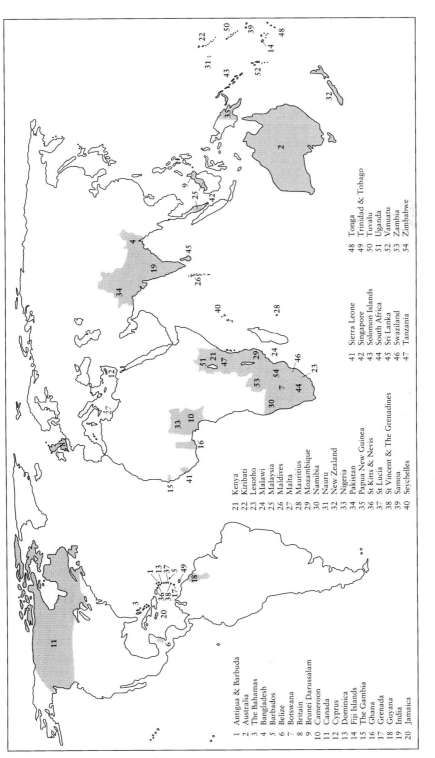

1 Antigua & Barbuda
2 Australia
3 The Bahamas
4 Bangladesh
5 Barbados
6 Belize
7 Botswana
8 Britain
9 Brunei Darussalam
10 Cameroon
11 Canada
12 Cyprus
13 Dominica
14 Fiji Islands
15 The Gambia
16 Ghana
17 Grenada
18 Guyana
19 India
20 Jamaica

21 Kenya
22 Kiribati
23 Lesotho
24 Malawi
25 Malaysia
26 Maldives
27 Malta
28 Mauritius
29 Mozambique
30 Namibia
31 Nauru
32 New Zealand
33 Nigeria
34 Pakistan
35 Papua New Guinea
36 St Kitts & Nevis
37 St Lucia
38 St Vincent & The Grenadines
39 Samoa
40 Seychelles

41 Sierra Leone
42 Singapore
43 Solomon Islands
44 South Africa
45 Sri Lanka
46 Swaziland
47 Tanzania

48 Tonga
49 Trinidad & Tobago
50 Tuvalu
51 Uganda
52 Vanuatu
53 Zambia
54 Zimbabwe

Figure 5.3 The Commonwealth

global capitalism/ International flows of people, capital investment, aid and debt repayments often continue to reflect past colonial ties. Different conflicts in the world often have colonial roots, as shown by the dispute between India and Pakistan over Kashmir and between nationalists and unionists in Northern Ireland. Direct colonial rule continues in many places, as shown by the Chinese occupation of Tibet and the Indonesian control of East Timor (Pilger, 1998). (

Box 5.3 The Commonwealth

In 1926, the Imperial Conference in London accepted the Balfour Report that described dominions such as Canada, Australia, New Zealand and South Africa as 'autonomous Communities within the British Empire, equal in status, in no way subordinate to one another ... united by a common allegiance to the Crown, and freely associated as members of the British Commonwealth of Nations' (quoted in Chamberlain, 1985: 53). But the modern Commonwealth is usually dated from 1949, when India was admitted as the first republican member and 'British' was dropped from the title of the association. While the original Commonwealth comprised white settler colonies and was referred to as a 'White Man's Club', the modern Commonwealth has 54 member states from Africa, the Caribbean, the Mediterranean and Asia and is committed to policies of racial equality and national sovereignty. Today, 33 member states are republics, five have their own national monarchies and 16 are constitutional monarchies that recognise Queen Elizabeth as their head of state (*Guardian*, 28 October 1997). In 1995, Mozambique was the first country with no colonial ties to Britain to join the Commonwealth. In the same year, Nigeria was suspended from the Commonwealth following the execution of Ken Saro-Wiwa and other human rights activists. Commonwealth Principles include international peace and order, individual liberty and international co-operation. Two other Principles are 'We recognise racial prejudice as a dangerous sickness' and 'We oppose all forms of colonial domination' (*Guardian*, 28 October 1997; also see Chamberlain, 1985, and Darwin, 1988; and the websites of the Royal Commonwealth Society and the Commonwealth Institute at *http://www.rcsint.org* and *commonwealth.org.uk*).

Imperial and colonial geographies continue to shape the world. While the political and economic basis and effects of colonialism should not be underestimated, the *cultural* basis and effects of colonialism have been attracting a great deal of critical attention in recent years. As Nicholas Dirks writes,

> Although colonial conquest was predicated on the power of superior arms, military organization, political power, and economic wealth, it was also based on a completely related variety of cultural technologies. Colonialism not only has had cultural effects that have too often been either ignored or displaced into the inexorable logics of modernization and world capitalism, it was itself a cultural project of control. Colonial knowledge both enabled colonial conquest and was produced by it; in certain important ways, culture was what colonialism was all about. (Dirks, 1992: 3)

For Dirks, colonialism was a cultural as well as a military, economic and political project of control over space, mobilising different webs of meaning, practices, objects of knowledge and ways of knowing that helped to legitimate and to perpetuate colonial power. But a focus on the cultural basis of colonial power does not efface the violence of conquest and control. Rather, an examination of the cultural 'structures of meaning' that underpinned colonial rule can reveal the ways in which the violence of conquest and control was exercised, justified and represented. According to Nicholas Thomas,

> colonialism is not best understood primarily as a political or economic relationship that is legitimized or justified through ideologies of racism or progress. Rather, colonialism has always, equally importantly and deeply, been a cultural process; its discoveries and trespasses are imagined and energized through signs, metaphors and narratives; even what would seem its purest moments of profit and violence have been mediated and enframed by structures of meaning. Colonial cultures are not simply ideologies that mask, mystify or rationalize forms of oppression that are external to them; they are also expressive and constitutive of colonial relationships in themselves.
> (Thomas, 1994: 2)

In his 'geographical inquiry into historical experience', Edward Said refers to imperial and colonial struggles in explicitly geographical terms: 'Just as none of us is outside or beyond geography, none of us is completely free from the struggle over geography'. As he continues, these geographical conflicts are not only military but also cultural: 'That struggle is complex and interesting because it is not only about soldiers and cannons but also about ideas, about forms, about images and imaginings' (Said, 1993: 7). Rather than separate the military, economic, political and cultural basis and effects of colonialism, it is important to emphasise their links and material consequences in particular contexts. So, the military might of soldiers and cannons cannot be isolated from the structures of meaning – ideas, imaginings, images and representations – that have helped to legitimate both conflict and resistance to such conflict (see Dawson, 1994, for a discussion of gendered representations of 'soldier heroes'). In recent years, such structures of meaning and cultural representations over space have been a central concern of colonial discourse analysis.

Colonial discourse analysis

'Colonial discourse' refers to the apparatus of power that legitimates colonial rule over people and places at a distance. 'Colonial discourse analysis' involves the interrogation of colonial power through the critical study of colonial discourses. Colonial discourse analysis challenges the ways in which colonial power and western knowledges become taken for granted and naturalised by questioning 'Western knowledge's categories and assumptions' (Young, 1990: 11). Colonial discourse analysis involves the study of colonial ideas, images and imaginings to explore the basis and effectiveness of colonial power and knowledge. In the words of Ania Loomba,

> 'Colonial discourse' ... is not just a fancy new term for colonialism; it indicates
> a new way of thinking in which cultural, intellectual, economic or political
> processes are seen to work together in the formation, perpetuation and
> dismantling of colonialism. It seeks to widen the scope of studies of
> colonialism by examining the intersection of ideas and institutions,
> knowledge and power. (Loomba, 1998: 54)

The postcolonial work of Edward Said, Homi Bhabha and Gayatri Chakra-vorty Spivak has been particularly important in widening the study of colonial ideas and institutions, knowledge and power.

'Orientalism': Edward Said

Edward Said's book, *Orientalism*, was first published in 1978 and represents a landmark in the field of colonial discourse analysis. Many critics credit *Orientalism* with inaugurating a new area of academic inquiry, claiming that it is 'commonly regarded as the catalyst and reference point for postcolonialism' (Gandhi, 1998: 64). Said's work should not be seen in isolation from earlier studies of colonial power and ways of knowing that include work by Fanon, Memmi and Césaire (see Box 5.4 for an introduction to Fanon's work and the Algerian War of Independence). But Said's work has inspired an unprecedented reorientation of research into the colonial politics of representation and identity across many disciplines.

Box 5.4 Frantz Fanon and the Algerian War of Independence

Frantz Fanon (1925–1961) was born in the French Caribbean colony of Martinique and trained as a psychiatrist in France. His work on racism and colonialism is most closely associated with the Algerian liberation movement and has been extremely influential in theorising an anti-colonial revolutionary consciousness. Fanon's best-known books are *Black Skin, White Masks* (1952), *A Dying Colonialism* (1959) and *The Wretched of the Earth* (1961). Fanon combined his clinical insights into the psychic effects of colonial domination with a marxist understanding of social and economic domination. As Ashcroft *et al.* explain, 'From this conjunction he developed his idea of a *comprador* class, or élite, who exchanged roles with the white colonial dominating class without engaging in any radical restructuring of society. The black skin of these compradors was "masked" by their complicity with the values of the white colonial powers. Fanon argued that the native intelligentsia must radically restructure the society on the firm foundation of the people and their values' (Ashcroft *et al.*, 1998: 99; also see Gordon *et al.*, 1996).

Fanon moved to Algeria in 1953. Algeria had been a French colony since the early nineteenth century and was the only part of the French empire that had attracted a significant number of French settlers (Chamberlain, 1985). The Revolutionary Committee for Unity and Action was formed by young nationalists in 1954 and 'divided the country into a number of districts (*wilayas*) each with a rebel leader at its head, a well-organised group of activists and whatever armaments could be procured. On 31 October, a CRUA declaration demanded independence and called Algerians to arms. In the early morning of

Box 5.4 *continued*

1 November, simultaneous insurrections broke out in each *wilaya*; although the French ignored the manifesto and not all of the attacks were successful, the Algerian war of independence had begun' (Aldrich, 1996: 292–293). This war lasted until Algeria gained independence on 3 July 1962. The CRUA developed into the National Liberation Front (*Front de Libération Nationale* or FLN). The Battle of Algiers (see Figure 5.4.1.) from September 1956 to October 1957 was 'the ugliest phase of the war' (Aldrich, 1996: 294), with the terrorist attacks of the FLN punished by French torture and brutality. By the end of the war, France estimated that 12,000 French soldiers, 227,000 Algerian soldiers and 20,000 civilians had been killed, while the FLN estimated that 1 million Algerian Muslims had been killed. More recent estimates suggest that there were half a million casualties (Aldrich, 1996).

Figure 5.4.1 The Battle of Algiers, 1956–57

During the war, 'Fanon supported the revolutionaries by utilizing hospital resources to train them in emergency medicine and psychological techniques for resisting torture, as well as fighting techniques' (Gordon *et al.*, 1996). In *A Dying Colonialism*, Fanon analysed the revolutionary transformations in Algerian consciousness over the course of the war. As he wrote, 'What we Algerians want is to discover the man behind the colonizer; this man who is both the organizer and the victim of a system that has choked him and reduced him to silence. As for us, we have long since rehabilitated the Algerian colonized man. We have wrenched the Algerian man from a centuries-old and implacable oppression. We have risen to our feet and we are now moving forward. Who can settle us back in servitude? We want an Algeria open to all, in which every kind of genius may grow' (Fanon, 1967: 32). Fanon concluded that 'The Revolution in depth, the true one, precisely because it changes man and renews society, has reached an advanced stage. This oxygen which creates and shapes a new humanity – this, too, is the Algerian Revolution' (Fanon, 1967: 181)

In *Orientalism*, Said examines the complex interactions of power, knowledge and representation. Geography lies at the heart of his analysis, as he focuses on the 'imaginative geographies' produced by the West about the East (Gregory, 1995a, 1995b). More specifically, Said discusses British, French and American imaginings of the Middle East over the last two centuries. Said is concerned to trace and to critique the orientalist discourses that imagined and, indeed, *produced* 'the Orient' as distinct from the western 'Occident'. As Said writes,

> The Orient was almost a European invention, and had been since antiquity a place of romance, exotic beings, haunting memories and landscapes, remarkable experiences .... The Orient is not only adjacent to Europe; it is also the place of Europe's greatest and richest and oldest colonies, the source of its civilizations and languages, its cultural contestant, and one of its deepest and most recurring images of the Other. In addition, the Orient has helped to define Europe (or the West) as its contrasting image, idea, personality, experience. Yet none of this Orient is merely imaginative. The Orient is an integral part of European *material* civilization and culture. (Said, 1978: 1–2)

As this quotation suggests, the Orient and the Occident were defined in relation to each other. While the Orient was represented as romantic, exotic, mysterious, and dangerous, the Occident was assumed to be the norm against which it was different or 'other'. In this way, Orientalism produced knowledges about colonised people and places as 'other', inferior and irrational in contrast to a powerful, rational, western 'self'. As Said stresses, imaginative geographies of *both* West and East were produced by orientalist discourses, suggesting the relational constitution of identity and a spatial politics of difference. As Gillian Rose puts it, 'The Orient was defined as exotic, decadent and corrupt, but these verdicts were passed in relation to a West which implicitly situated itself as civilized and moral in contrast .... What were seen as Oriental vices were made to define Western European virtues' (Rose, 1995: 93). Imaginative geographies of the Orient came to be seen as the *reality* of the Middle East. But rather than reveal the reality of life in the Middle East, such Orientalist representations revealed more about western life, culture, fears and desires: 'Orientalism ... has less to do with the Orient than it does with "our" world' (Said, 1978: 12).

Said identifies three, interrelated meanings of Orientalism. First, he points to the scholarly tradition of Orientalism that came to prominence in post-Enlightenment Europe and encompassed work in literature, history, philology and archaeology. European academics produced a wealth of research on the Middle East at this time, extending a fascination with the Orient that stemmed from before the Crusades. Second, Said refers to Orientalism in a more general sense, away from the libraries and colleges of academic research. Here, he refers to more widely diffused perceptions of the differences between West and East that were produced through, for example, travel writing, art and literature. Crucially, Said points to the constant interchange between these Orientalist knowledges, within but also beyond the academy. Finally, Said highlights the status of Orientalism as 'a Western style for dominating, restructuring, and having authority over the Orient' (Said, 1978: 3), demonstrating the ways in which imaginative geographies of the Orient underpinned imperial and colonial rule. The prevalence and authority of Orientalist ways of seeing, thinking and knowing in the West 'justified colonialism in advance as well as subsequently facilitating its successful operation' (Young, 1990: 129). These three, connected meanings of Orientalism reveal the imbrication of power and knowledge, the ways in which Orientalist representations came to be seen as reality, and their material effects in shaping, maintaining and justifying imperial and colonial geographies. For Said,

> without examining Orientalism as a discourse one cannot possibly understand the enormously systematic discipline by which European culture was able to manage – and even produce – the Orient politically, sociologically, militarily, ideologically, scientifically, and imaginatively during the post-Enlightenment period. (Said, 1978: 3)

It is important to remember that *Orientalism* is 'a book not about non-Western cultures, but about the Western representation of these cultures' (Loomba, 1998: 43). But this distinction was blurred within Orientalist discourses as western representations of non-western cultures were seen to reflect reality. According to Said, 'Orientalism was ultimately a *political* vision of reality whose structure promoted the difference between the familiar (Europe, the West, "us") and the strange (the Orient, the East, "them")' (Said, 1978: 43). This political vision of reality was an imperial and a colonial vision.

Edward Said's work on Orientalism has inspired an enormous amount of research into the colonial politics of representation, the intersections of power and knowledge, and identity formation over space and time. Such research has focused not only on past Orientalist discourses but also on the persistence of similar discourses and strategies of 'othering' in the world today. Just as Said explored the imaginative geographies produced by the West about the East, a similar process can be seen in First World stereotypes about the 'Third World' as less advanced and powerless (Bell, 1994). Many people and places beyond the West continue to be represented as 'exotic', suggesting that the West is the norm against which 'others' are defined as different. So, for example, tourist brochures often represent people and places as exotic and 'other' to western life, as shown in Figure 5.4, in which the white woman seems to embody the norm against which the Long Neck Paduang woman in Thailand appears to be exotically different (also see Cook, 1993; McClintock, 1995; and Richards, 1990, for parallel examples in the advertising and consumption of commodities such as bananas and soap and the emergence of 'commodity racism'). Both in the past and in the present, western desire for the 'other' is often explicitly sexualised (see Chapter 4). In the nineteenth century, written and visual representations of Middle Eastern harems (Kabbani, 1986) revealed much about the sexual fantasies of many western male travellers that could be more easily played out away from the confines of life at home: 'Europe was charmed by an Orient that shimmered with possibilities, that promised a sexual space, a voyage away from the self, an escape from the dictates of the bourgeois morality of the metropolis' (Kabbani, 1986: 67). In a similar way, the growth of

un, Bangkok at sunset

Departures - Every Sunday by Royal Orchid service of Thai International.
(Also available daily on request/supplement).
Meals - As outlined *(B), (L)*   **Sightseeing** - Included on days 5-9 as outlined.
Alternative Hotels - As featured in the brochure on request/supplement.

Long Neck Paduang, Mae Hong Sorn

Beachside, Phuket

PAGE 103

Figure 5.4 Representing Thailand to tourists

international sex tourism in Southeast Asia since the Second World War (Enloe, 1990; Law, 1997; Lee, 1991) owes much to stereotypical representations of 'Oriental' women as more sensuous and sexually available than their western counterparts. Just as Said argued in *Orientalism*, these representations say more about the West than about an Oriental reality and they have material consequences in the form of violence, exploitation and dependence.

A number of critiques of *Orientalism* have also been important in stimulating postcolonial studies across a range of disciplines (see Mani and Frankenberg, 1985, for an overview of these critiques and Said's response to them in Said, 1985). Said has been criticised for perpetuating the distinction between Orient and Occident that he seeks to challenge and for representing this distinction in terms that are too static and general. Moreover, Said concentrates on western representations of the Orient and pays little attention to self-representations of the colonised and strategies of resistance (but see Said, 1993, where he discusses resistance in greater detail). At the same time, several critics have argued that Said's representation of Orientalism is too internally coherent, obscuring the fissures, contradictions and vulnerabilities of imperial and colonial power and knowledge. Building on Said's work, Lisa Lowe writes that multiple 'orientalist situations' existed at different times and in different places and that each of these was internally complex, unstable and contradictory (Lowe, 1991). She argues that Orientalism was one among many discourses that included 'the medical and anthropological classifications of race, psychoanalytic versions of sexuality, or capitalist and Marxist constructions of class' (Lowe, 1991: 8). Other commentators have shown that the contradictions at the heart of imperial and colonial rule destabilised its apparent hegemony. The exposure of such internal contradictions can begin to disrupt an essentialist, unitary view of imperial and colonial power as 'a coherent imposition, rather than a practically mediated relation' (Thomas, 1994: 3). As Nicholas Thomas writes,

> Colonizing projects were ... frequently split between assimilationist and segregationist ways of dealing with indigenous peoples; between impulses to define new lands as vacant spaces for European achievement, and a will to define, collect and map the cultures which already possessed them; and in the definition of colonizers' identities, which had to reconcile the civility and values of home with the raw novelty of sites of settlement. (Thomas, 1994: 2–3)

Furthermore, Orientalist discourses were also profoundly gendered, which is largely overlooked by Said. An increasing amount of feminist work has revealed the gendered basis of colonial discourses such as Orientalism and the different roles played by men and women as both colonisers and colonised (Blunt and Rose, 1994; Chaudhuri and Strobel, 1992; Melman, 1992; Midgley, 1998; see Chapter 3 for more on the links between imperial and feminist politics).

The next two sections introduce the work of Homi Bhabha and Gayatri Chakravorty Spivak. While Bhabha turns his attention to the creation of colonial subjects through a focus on ambivalence, mimicry and hybridity, Spivak focuses on the gendered as well as racialised domination that characterised imperial and colonial rule and examines the difficulties of recovering the agency and experiences of colonised people.

Ambivalent positions: Homi Bhabha

While *Orientalism* concentrated on the production of colonial discourses, the work of Homi Bhabha has examined the place of colonised people in these discourses. Unlike Said's focus on the hegemonic power of Orientalism, Bhabha turns to the fractures and ambivalences that destabilised imperial and colonial rule, stresses the relational basis and instabilities of such rule, and challenges the polarities of West and East, 'self' and 'other' (Bhabha, 1994). Inspired by the work of Fanon, Bhabha invokes the psychoanalytic term 'ambivalence' to convey 'the complex mix of attraction and repulsion that characterizes the relationship between colonizer and colonized' (Ashcroft *et al.*, 1998: 12). Rather than represent the colonised subject as simply *either* complicit *or* opposed to the coloniser, Bhabha suggests the *coexistence* of complicity and resistance. The hegemonic authority of colonial power is made uncertain and unstable because the ambivalent relationships between colonisers and colonised are complex and contradictory. Bhabha discusses the creation and efficacy of colonial stereotypes in terms of ambivalence. For Bhabha, a stereotype is profoundly ambivalent because it 'is a form of knowledge and identification that vacillates between what is always "in place", already known, and something that must be anxiously repeated' (Bhabha, 1994: 66). It is precisely this notion of excess beyond the control of the coloniser that reveals the limits and vulnerabilities of colonial authority. As Bhabha suggests,

> the colonial stereotype is a complex, ambivalent, contradictory mode of representation, as anxious as it is assertive, and demands not only that we extend our critical and political objectives but that we change the object of analysis itself. (Bhabha, 1994: 70)

For Bhabha, the study of colonial discourse should address the creation of colonial subjects, moving beyond the identification of images as positive or negative to a more detailed analysis of the processes of subjectification. While colonial discourse might appear to fix difference in its representations of 'the other', Bhabha stresses the ambivalent nature of this apparent fixity. For Bhabha, 'because the colonial relationship is always ambivalent, it generates the seeds of its own destruction' (Ashcroft et al., 1998: 13).

Connected with his ideas about the excess of colonial stereotypes, Bhabha also represents the ambivalent relationships between colonisers and colonised in terms of mimicry. This term refers to the ways in which colonised subjects adopt the coloniser's habits, lifestyle and values. And yet, this adoption is never a straightforward reproduction and 'mimicry is never very far from mockery, since it can appear to parody whatever it mimics' (Ashcroft et al., 1998: 139). Bhabha cites the example of an Indian man in the nineteenth century, educated in English and working in the Indian Civil Service, who moves within and between the lives of colonisers and colonised. As Young explains,

> The mimic man, insofar as he is not entirely like the colonizer, white but not quite, constitutes only a partial representation of him: far from being reassured, the colonizer sees a grotesquely displaced image of himself. Thus the familiar, transported to distant parts, becomes uncannily transformed, the imitation subverts the identity of that which is being represented, and the relation of power, if not altogether reversed, certainly begins to vacillate. (Young, 1990: 147)

The effect of such mimicry is to undermine colonial authority by exposing 'a crack in the certainty of colonial dominance, an uncertainty in its control of the behaviour of the colonized' (Ashcroft et al., 1998: 139) and by revealing 'colonialism's own vulnerability to itself' (Jacobs, 1996: 26).

The relational identities of coloniser and colonised and the instabilities of colonial rule are also revealed in Bhabha's notions of hybridity. For Bhabha, colonial domination results in a process of transcultural hybridisation, whereby 'certain elements of dominant cultures are appropriated by the colonised and rearticulated in subversive ways' (Jacobs, 1996: 28). The hybrid subject is a split and a mobile subject, reflecting the flux between colonising and colonised identities. Bhabha locates hybridity in explicitly spatial terms, referring to a contradictory and ambivalent 'Third Space' that disrupts the binary opposition between 'self' and 'other'. For Bhabha, 'Third Space' is an in-between space – within and between the fissures of colonial rule – where resistance can be enunciated and where hierarchies between cultures, colonisers and colonised become untenable. As a result, hybridity becomes an empowering way to envisage cultural difference that contrasts with representations of an exotic and usually inferior 'other' to a western 'self'. Bhabha writes that travelling into Third Space

> may open the way to conceptualizing an *inter*national culture, based not
> on the exoticism of multiculturalism or the *diversity* of cultures, but on the
> inscription and articulation of culture's *hybridity*. To that end we should
> remember that it is the 'inter' – the cutting edge of translation and negotiation,
> the *in-between* space – that carries the burden of the meaning of culture.
> (Bhabha, 1994: 38)

Hybrid meanings of culture overturn ideas about cultural, national and racial 'purity' by examining processes of transculturation (Bhabha, 1994; Pratt, 1992; Young, 1995). As Mary Louise Pratt shows in her analysis of European imperial travel writing in South and Central America, transculturation is 'a phenomenon of the contact zone' where colonisers and colonised interact (Pratt, 1992). But this interaction reflects and reproduces colonial power relations. As Pratt explains, 'contact zones' are places where 'disparate cultures meet, clash and grapple with each other, often in highly asymmetrical relations of dominance and subordination – like colonialism, slavery, or their aftermaths as they are lived out across the globe today' (Pratt, 1992: 4). Despite such asymmetrical power relations, processes of transculturation are not just one-way:

> While the imperial metropolis tends to understand itself as determining the
> periphery (in the emanating glow of the civilizing mission or the cash flow of
> development), it habitually blinds itself to the ways in which the periphery
> determines the metropolis, beginning, perhaps, with the latter's obsessive need
> to present and represent its peripheries and its others continually to itself.
> (Pratt, 1992: 6)

For Bhabha, 'the location of culture' is in the in-between, ambivalent spaces of contact between colonisers and colonised. As a result, an Orientalist binary of 'self' and 'other' becomes more complicated, and nationalist claims to a bounded, exclusive identity become unsustainable. Bhabha's analysis of colonial discourse is clearly geographical, relating to the simultaneous distancing and interchange of colonisers and colonised both on an unconscious level and in colonial contact zones. And yet, many critics have criticised the lack of geographical and historical specificity in Bhabha's work. As Loomba writes, 'Bhabha generalises and universalises the colonial encounter' and, ironically, 'the split, ambivalent, hybrid colonial subject projected in his work is in fact curiously universal and homogeneous – that is to say he could exist anywhere in the colonial world' (Loomba, 1998: 178). Loomba's gendered language is deliberate here, reflecting the lack of attention paid by Bhabha to gender difference in colonial encounters. Moreover, Bhabha has also been criticised not only for employing the western language of psychoanalysis in universalising ways, but also for overlooking the gendered and sexual implications of psychoanalytic enquiry (Young, 1990; see Chapter 4 for an introduction to Freudian and Lacanian psychoanalysis). In the next section, the gendered nature of colonial discourse will be introduced through a discussion of Gayatri Chakravorty Spivak's work.

'Can the subaltern speak?': Gayatri Chakravorty Spivak

Gayatri Chakravorty Spivak's work is diverse and complicated (Landry and MacLean, 1996; Spivak, 1990; and Young, 1990, provide good starting points). This section will focus on a well-known question first posed and answered by Spivak in 1985: 'can the subaltern speak?' (Spivak's essay is reprinted in Williams and Chrisman, 1993: 66–111). Returning to the quotation from Jonathan Crush at the beginning of the chapter, this question asks whether the colonial underclass can articulate its experiences and be active, visible and audible subjects. In geographical terms, Spivak's question relates to the 'hidden spaces' occupied by the colonial underclass and whether it is possible to speak from such locations. Spivak is concerned with the difficulties of recovering the voices of colonised people and, in particular, the voices of colonised women. Spivak concludes that 'the subaltern cannot speak', for reasons that will be explored below.

The term 'subaltern' has a military origin, referring to inferior ranks below the officer class. Its meaning was broadened by Antonio Gramsci, who used it to refer to groups such as peasants and workers who are subordinate to the hegemonic power of the ruling classes and who sought to write histories from below about the experiences of subaltern groups (Ashcroft et al., 1998; see also Chapter 2). More recently, the term has been associated with the work of the Subaltern Studies collective. Over the course of the 1980s, this collective of historians published the first five volumes of essays relating to subalternity in South Asia, a term which they use to refer to 'the general attribute of subordination in South Asian society whether this is expressed in terms of class, caste, age, gender and office or in any other way' (Guha, 1982: vii; see Guha and Spivak, 1988, for a collection of essays from these five volumes). The histories written by the Subaltern Studies collective seek to redress the élitism of imperialist and nationalist histories. While imperialist histories often focused on an imperial élite, nationalist histories often celebrated a bourgeois, nationalist élite, neither of which left space for the experiences of subordinate subalterns and a 'politics of the people' (Guha, 1982: 4). The Subaltern Studies collective aimed to write new, critical histories of South Asia, which followed the traditions neither of imperialist histories of conquest, nor of nationalist histories that charted a singular and linear development of nationalist consciousness. Subaltern histories focused on the lives, agency and resistance of those people who had been silenced and erased from both imperialist and nationalist accounts of the past.

By posing the question 'can the subaltern speak?', Spivak challenges the assumption that it is possible to recover subaltern voices. Inspired by marxism, deconstructionism and feminism, Spivak focuses on the place of Indian women as silenced subalterns, arguing that they were doubly colonised as Indian but also as women. As Kabbani puts it:

Eastern women were doubly inferior, being women *and* Easterners. They were an even more conspicuous commodity than their Western sisters. They were part of the goods of empire, the living rewards that white men could, if they wished to, reap. (Kabbani, 1986: 51)

For Spivak, if 'in the context of colonial production, the subaltern has no history and cannot speak, the subaltern as female is even more deeply in shadow' (Spivak, 1993: 83).

Spivak argues that the subaltern cannot speak because she has been silenced and objectified by colonial power and erased from imperialist and nationalist histories. For Spivak, subaltern spaces remain hidden from view. She contends that it is a fraught and ultimately impossible task to uncover these hidden spaces and to hear subaltern voices. The task, Spivak suggests, is one of examining the operation of power that has so effectively silenced and objectified the subaltern. Rather than repeating such an operation of power through silencing and objectification, Spivak argues that it is necessary to unlearn privilege and to decolonise gender. Ultimately, the aim should be to speak *with* rather than to speak *for* or merely try to listen *to* the historically mute subaltern subject.

Spivak develops her analysis by examining discursive representations of *sati*. Sati refers to the practice of Hindu widow sacrifice, when a Hindu widow climbs onto the funeral pyre of her husband to burn with his body. Sati was outlawed by British rulers in India in 1829 but did not completely disappear. Since Independence in 1947, there have been about 40 reported cases of sati and there has been a marked increase over the last decade. While a number of attempts have been prevented by the police, about five have been successfully carried out and have been accompanied by widespread media coverage in both India and abroad (Sunder Rajan, 1993). While there have been campaigns against sati by a range of groups, particularly women's organisations, there have also been public displays that support and glorify the practice. Representing sati is a complicated issue as it is located within a nexus of complex and competing discourses (Mani, 1987, 1992; Sunder Rajan, 1993; Spivak, 1993). For Spivak, the figure of the Hindu widow committing sati embodies the subaltern subject who is unable to speak.

Spivak examines colonial and nativist representations of sati, arguing that while these views were polarised, they both served to silence the subaltern subject. British colonial rulers represented sati as a forced, primitive and barbaric act that needed to be prevented: 'she must be saved from dying'. But, as Spivak suggests, a material consequence of these representations was to extend and to strengthen British colonial power by legitimating a paternal colonialism whereby 'white men are saving brown women from brown men' (Spivak, 1993: 92). In nativist accounts, celebrating pre-colonial life and traditions, sati was represented as a voluntary practice, displaying the loyalty and devotion of a Hindu widow to her late husband: 'she wanted to die'. Spivak argues that these polarised views, representing sati as either forced or voluntary, both served to silence the Hindu widow and to erase her from view. In colonial discourses, the subaltern subject was defined only in relation to her husband and existed only to be saved from the barbarism of other Indians. In nativist discourses, the subaltern subject was still defined in terms of her husband and became visible only in the pain and the spectacle of her death which itself took place only because of her husband's death. In both colonial and nativist discourses, the subject of sati – the figure of the Hindu widow – is objectified as a victim who is voiceless and powerless and lacks agency of her own.

Challenging the assumption that subaltern voices can be recovered independently from colonial and nativist discourses, Spivak writes that the Hindu widow 'cannot speak'. There are very few records of the feelings and experiences of women themselves who committed sati. But more than this, Spivak argues that colonial and nativist discourses made it impossible for such women to speak and to be heard. By posing the question 'can the subaltern speak?' Spivak critiques the assumption that a unified subaltern identity can be recovered that is autonomous from colonial and nativist discourses. Her declaration that 'the subaltern cannot speak'

> has sometimes been interpreted to mean that there is no way in which oppressed or politically marginalized groups can voice their resistance, or that the subaltern only has a dominant language or a dominant voice in which to be heard. But Spivak's target is the concept of the subaltern subject's ability to give voice to political concerns. Her point is that no act of dissent or resistance occurs on behalf of an essential subaltern subject entirely separate from the dominant discourse that provides the language and the conceptual categories with which the subaltern voice speaks. (Ashcroft *et al.*, 1998: 219)

For Spivak, the subaltern is both created and silenced by dominant discourses.

Spivak's analysis destabilises attempts simply to recover subaltern histories and voices and points to the importance of decolonising ideas about gender. Clear parallels exist between her work on the objectification and silencing of the Hindu widow and other work that critiques the stereotypical ways in which women from the 'Third World' often continue to be represented (see, for example, the essays in Mohanty *et al.*, 1991, and in Alexander and Mohanty, 1997). Writers such as Chandra Talpade Mohanty and Aihwa Ong critique the ethnocentric representations of non-western women by many western feminists and highlight the ways in which, for many women, feminism represents a white, western and bourgeois form of cultural imperialism, as discussed in Chapter 3. The diverse experiences and agency of non-western women are often reduced to a singular, passive and static category of 'Third World woman'. As Ong writes,

> The status of non-Western women is analyzed and gauged according to a set of legal, political, and social benchmarks that Western feminists consider critical in achieving a power balance between men and women .... By portraying women in non-Western societies as identical and interchangeable, and more exploited than women in the dominant capitalist societies, liberal and socialist feminists alike encode a belief in their own cultural superiority. (Ong, 1988: 82, 85)

Such ethnocentric representations are imperial because they assume that western standards and feelings are superior to those elsewhere in the world. The status and roles of non-western women are often reported in sensationalist terms or are otherwise reduced to different indicators of development, losing 'the everyday, fluid, fundamentally historical and dynamic nature of the lives of third world women' (Mohanty, 1988: 6). Such stereotypical representations mean that non-western women are often assumed to have the same problems, needs, goals and interests. But, as Mohanty illustrates,

> the interests of urban, middle-class, educated Egyptian housewives, to take only one instance, could surely not be seen as being the same as those of their uneducated, poor maids. Development policies do not affect both groups of women in the same way. Practices which characterize women's status and roles vary according to class. Women are constituted as women through the complex interaction between class, culture, religion, and other ideological institutions and frameworks. (Mohanty, 1988: 63)

The lives of non-western women are as diverse as the lives of their western counterparts.

The work of Said, Bhabha and Spivak has been very influential in shaping the diverse field of postcolonial studies. As these introductions to their work have shown, geography lies at the heart of their analyses. Colonial discourses were also geographical discourses, representing difference over space and legitimating territorial expansion and control. Geographical imaginations of 'other' people and places were inseparably bound up with imperial and colonial imaginations of the world. While Said traverses the imaginative geographies of Orientalist discourses, Bhabha proposes a Third Space of hybridity that destabilises a binary and static distinction between 'Self' and 'Other'. Finally, Spivak turns to the difficulties of recovering subaltern experiences and voices from their hidden spaces of confinement and objectification. The next section considers postcolonial geographies in more disciplinary terms by examining attempts to write critical histories of geography and imperialism and to represent the world today.

Postcolonial geographies

This section will begin by focusing on the history of geography as an academic discipline, exploring the connections between geographical knowledge, education and empire. More specifically, this section will consider the institutionalisation and increasing influence of British geography as it developed alongside British imperial and colonial policies of conquest and control in the late nineteenth century. At this time, geographical knowledge was intimately tied to imperial and colonial power. But, unlike other disciplines such as anthropology, it is only recently that geographers have started to write about the imperial complicity of the discipline, and to write about its imperial past in critical and contextual ways rather than celebrating exploration, 'discovery', and heroic explorers and geographers. It can be argued that 'geography was the science of imperialism *par excellence*' because 'exploration, topographic and social survey, cartographic representation, and regional inventory ... were entirely suited to the colonial project' (Livingstone, 1993: 160, 170). But because imperialism and colonialism were about far more than economic exploitation alone, Felix Driver has called for critical, contextual histories of geography that examine 'the *culture* of imperialism' (Driver, 1992). Geography as an academic discipline and geographical education at all levels played fundamental roles in shaping the ideas, meanings and imaginations that helped to represent and to justify the British Empire (see Maddrell, 1998, and Ploszajska, 1998, for more on geographical education in British schools in the nineteenth and early twentieth centuries).

The Royal Geographical Society was the main institution for the development and promotion of geographical research and education in Britain in the nineteenth century. From about 1870, an emerging 'new geography' was characterised by increasing professionalisation and the movement to establish the subject as a formal academic discipline (Stoddart, 1986). By the 1880s and 1890s, though, the increased specialisation of other disciplines such as geology meant that although British geography was emerging as a scientific discipline, institutions such as the RGS lacked 'strictly scientific men' (Stoddart, 1986: 67). Calls for geographical education were often explicitly tied to imperial and colonial imperatives at a time when the British Empire was at its height. This is vividly shown in a report by John Scott Keltie on the state of geographical education in Britain in 1886:

> There is no country that can less afford to dispense with geographical knowledge than England ... [yet] there are few countries in which a high order of geographical teaching is so little encouraged. The interests of England are wide as the world. Her colonies, her commerce, her emigrations, her wars, her missionaries, and her scientific explorers bring her into contact with all parts of the globe, and it is therefore a matter of *imperial importance* that no reasonable means should be neglected of training her youth in sound geographical knowledge. (Keltie, quoted in Blunt, 1994: 8; emphasis added)

For Keltie, geographical knowledge and imperial power were closely entwined and he believed that 'sound geographical knowledge' was vital for the success of the British Empire.

The most tangible relationship between imperial power and geographical knowledge was through exploration and so-called discovery, charting territory that was often thought to be unknown and mapping and naming apparently 'new' places. Of course, such places were only unknown and 'new' to their European rulers, who often represented vast areas as unoccupied, blank spaces brought into being only under a European gaze. As Box 5.5 shows, mapping previously unknown places came to *create* such places in an imperial imagination that was also quite clearly a *geographical* imagination. Maps enabled European rulers to exercise power and control over territories elsewhere in the world. The exercise of such power and control over people as well as places was supported by other techniques of 'knowing' the world, which were also believed to be scientifically based and objective. So, for example, the natural world came to be named and classified in a systematised way, which ordered and controlled the diversity of species in apparently scientific ways (Pratt, 1992). But such ideas about classification and the standardised production of information and knowledge were not limited to the natural world of flora and fauna. Rather, imperial geography also contributed to the widespread attempt to survey and classify *people* throughout the world. Colonised people were objectified through, for example, measuring and classifying anatomical difference and photographing different racial 'types' (McClintock, 1995; Ryan, 1997). Geographical methods of survey, classification and measurement were about not only knowing and controlling physical environments, but also knowing and controlling people. Geography was a tool of empire, lending apparently scientific credibility to imperial and racist ideologies. People as well as places were produced and represented as 'other' by imperial geography.

Box 5.5 Mapping places

Throughout history, maps have represented and shaped geographical knowledge about the world. Although maps may appear to reflect reality, locating factual information in increasingly scientific ways, many critics argue that all maps are socially constructed forms of knowledge. Rather than represent the world 'as it really is', maps are always partial and infused with different meanings. Maps are never unmediated representations of the world and even the most apparently scientific and technologically advanced map is not value-free. A map's *modes of representation* (such as its scale, projection, typography and use of colour) and *what it represents* (the features that are included and excluded, the boundaries that delimit a map's area) both reveal the intimate connections between power and geographical knowledge (Harley, 1988). Brian Harley claims that 'maps are preeminently a language of power, not of protest' (Harley, 1988: 301) and shows that imperial maps were inextricably bound up with imperial power. Imperial maps were tools of empire: 'As much as guns and warships, maps have been the weapons of imperialism' (Harley, 1988: 282). Maps were used to chart and to claim new colonies, both representing and legitimating territorial expansion. Maps also helped to shape imaginative geographies of empire by representing territories as 'blank spaces' over which imperial rule could be imposed. The imperial gaze that was codified and reproduced by imperial maps produced places for conquest and control. In a similar way, strategies of naming brought places into existence in an imperial imagination that was also a geographical imagination (Carter, 1987; Huggan, 1989).

Many attempts have been made to decolonise maps and mapping. While the complicity of imperial maps with imperial power has been explored, so too have alternative, postcolonial mappings (see, for example, Blunt and Rose, 1994; Huggan, 1989; Jacobs, 1996; Nash, 1994; Phillips, 1997). For example, Jane Jacobs has written about the design of an Aboriginal art trail in Brisbane and the ways in which 'the conceptual template for the trail is based upon the creative appropriation of the map, that over-determined signifier of colonialism'

Box 5.5 *continued*

> (Jacobs, 1996: 144). Located at J. C. Slaughter Falls on the slopes of Mount
> Coot-tha, the art trail is part of a community arts project that seeks 'to restore
> Aboriginal cultural production and articulation to city space' (Jacobs, 1996:
> 144). An Aboriginalised map of the art trail is juxtaposed with a topographic
> map of the same place, and different mappings also appear in the artwork itself.
> Jacobs suggests that this creative reappropriation of mapping – this example of
> 'counter-cartography' – parodies colonial maps and destabilises the power
> inherent within them.

Imperial geography was also distinctively gendered, as shown by representations of imperial travel on which the production of geographical knowledge was seen to depend (Box 5.6). Both imperialism and geography were represented as masculine endeavours whereby imperial power and the geographical knowledge that helped to legitimate and to perpetuate such power were seen as masculine domains. For example, in 1887, the President of the RGS stated that:

> What we require ... is precise and accurate information of the earth's surface,
> however it may be obtained, and to train the minds of our youth in the powers
> of observation sufficient to enable them to obtain this information; if in so
> doing our countrymen continue to be stimulated to deeds of daring, to
> enterprise and adventures, to self-denial and hardships, it will assist in
> preserving the manhood of our country, which is more and more endangered
> year by year in consequence of our endeavour to keep peace within our own
> borders and to stave off strife with our neighbours. (Strachey, quoted in Blunt,
> 1994: 150)

The production of imperial geographical knowledge advocated by Strachey was embodied in exclusively masculine terms of virility and bravery. Moreover, Strachey describes this imperial masculinity as being under threat of emasculation at a time of peace. As Felix Driver contends, 'contemporary writings on "geography" were infused with assumptions about gender, as well as empire' (Driver, 1992: 28). These gendered assumptions often invoked a military as well as a scientific masculinity as geographical knowledge was closely tied to imperial conquest.

Box 5.6 **Postcolonial travels**

> Ideas about travel and travel writing have been important in postcolonial
> studies. In recent years, an increasing amount of critical attention has been paid
> to the ways in which travel is bound up with the production of knowledge,
> power relations and identity formations. The imperial development of
> disciplines such as geography and anthropology was closely linked to imperial
> travel and the immediacy and authority of observation associated with 'being

Box 5.6 *continued*

there' and viewing 'other' people and places at first hand. From the mid-nineteenth century, imperial travels were increasingly tied to the production of imperial knowledge and both gained scientific credibility and status. For example, a famous book published by the Royal Geographical Society stressed that 'it is the duty of every civilised traveller in countries newly opened up to research, to collect facts, plain unvarnished facts, for the information of those leading minds of the age who, by dint of great experience, can ably generalise from the details contributed from diverse sources' (Freshfield and Wharton, 1893: 446). Advice books for prospective travellers differed for men and women. While books written for male travellers focused on methods and equipment necessary for scientific observation, books written for female travellers often emphasised the appropriate appearance and behaviour of the traveller herself (Blunt, 1994). Imperial travel was clearly a gendered practice. Several writers have traced discourses of heroic masculinity that shaped imperial explorations (Driver, 1992; Pratt, 1992). Others have examined the ambivalent place of white women such as Mary Kingsley and Isabella Bird who were empowered to travel and to transgress in the empire while away from the feminised domesticity of living at home (Blunt, 1994; Mills, 1991). But for many people, travel is forced rather than voluntary, as shown by the displacement, exile and alienation of slaves, refugees, asylum seekers and the dispossessed (Bammer, 1994; Clifford, 1997; Hyndman, 1998).

Ideas about travel remain significant in postcolonial geographies of the present. Said has interpreted the production of knowledge over space and time in terms of 'travelling theory' (Said, 1983) and has written about marginalised subjectivities in terms of mobility:

> liberation as an intellectual mission ... has now shifted from the settled, established, and domesticated dynamics of culture, to its unhoused, decentred, and exilic energies, energies whose incarnation today is the migrant, and whose consciousness is that of the intellectual and artist in exile, the political figure between domains, between forms, between homes, and between languages. (Said, 1993: 332)

Bhabha also writes about the 'unhomely displacement' of the modern world (Bhabha, 1992), while Paul Carter suggests that 'an authentically migrant perspective ... might begin by regarding movement, not as an awkward interval between fixed points of departure and arrival, but as a mode of being in the world' (Carter, 1992: 101). Rather than celebrate travel in uncritical and ungrounded terms, it is important to explore diverse experiences of travel. James Clifford proposes the use of the term 'travel' 'precisely because of its historical taintedness, its associations with gendered, racial bodies, class privilege, specific means of conveyance, beaten paths, agents, frontiers, documents, and the like' (Clifford, 1992: 110).

Postcolonial geographies include work on the exclusions of the western academy (Crush, 1994; Robinson, 1994); written and visual representations over space (Blunt, 1994; Gregory, 1999; Phillips, 1997; Ryan, 1997); and interpreting urban landscapes (Jacobs, 1996; Mitchell, 1997). Reflecting the importance of geographical and historical specificity in postcolonial studies, this chapter will end by considering one example of postcolonial geographies of the present in some detail: the Great Famine Project organised by Action From Ireland.

'Famine is a lie': Postcolonial ideas and practice

Action From Ireland, or AFrI, was founded in 1975 and is an organisation working for justice, peace and human rights throughout the world. In 1988, AFrI launched the 'Great Famine Project' to commemorate the 150th anniversary of the Irish Famine (see Box 5.7) and to link these past experiences in Ireland to contemporary issues of famine, injustice and human rights abuses elsewhere in the world. As Joe Murray from AFrI explains:

> The aim of AFrI's 'Great Famine Project' is to ensure that this anniversary will not be allowed to slip by unnoticed, as happened on the 100th anniversary. We see the need for ourselves as a people to revisit this experience, to remember it and to deal with the trauma and the pain which it has left on the Irish psyche. But even more importantly our aim is to commemorate it in a way that motivates people to address the injustices and the inequalities that continue to create similar conditions for millions of people throughout the world today. (Murray, 1995: 3)

Box 5.7 The Irish Famine, 1845–50

It has been estimated that one million people died and one million people emigrated as a result of the Irish Famine (see, for example, Gray, 1995; Kelleher, 1997; Neal, 1997). As in other European countries, the potato crop was decimated by a blight in successive years during the 1840s. And yet, this was only the trigger rather than the cause of the Irish Famine. The causes of the Irish Famine – and the reasons why famine occurred in Ireland and not elsewhere in Europe – span a range of economic, social and political factors. In 1845, 4000 people out of a population of 8 million owned 80% of the land in Ireland. Many poor people rented land and grew cash crops to pay their landlords. These people were very often dependent on the potato crop for their own survival, with a working man eating up to 6.5 kg of potatoes each day. Many of those who owned land in Ireland were absentee landlords, often living in Britain because, at this time, Ireland was part of the British Empire and was ruled from London. Ireland was Britain's first and longest-standing colony. From as early as 1171, native kings in Ireland acknowledged Henry II of England as lord of Ireland. British colonial rule intensified from the sixteenth century onwards, and was marked by plantation agricultural systems and British settlement. The Act of Union in 1800 bound Ireland to Britain more closely than ever before. By the mid-nineteenth century, just as racist stereotypes characterised many British representations of colonised peoples in Asia and Africa, the appearance and behaviour of Irish people were also represented in racist terms. So, for

Box 5.7 *continued*

example, the Irish were caricatured as dirty, lazy and degenerate 'bogtrotters' who lived in squalor of their own making (McClintock, 1995). The potato crop failed in the context of social, economic and political polarisations between landlords, tenant farmers and landless labourers, and between imperial rulers and ruled.

During the famine, an estimated one million people died of starvation and diseases such as typhoid, cholera and dysentery. The British imperial response to the famine took the form of minimal state intervention. Cash crops continued to be exported from Ireland while people were dying of starvation, in line with the imperial policy of free trade. For many families, the only source of relief was to enter workhouses or participate in public works schemes, often building roads to nowhere in exchange for food. Thousands of families were evicted from their homes and an estimated one million people emigrated from Ireland in overcrowded ships where disease was rife. Most of those who emigrated travelled to the east coast of the United States; others settled in British ports such as Liverpool, while others travelled to eastern Canada and Australia. In Ireland, the famine led to popular dissent, such as food riots and the rise of Irish nationalism in the form of groups such as Young Ireland.

The legacy of the Irish famine is still felt today. In Ireland, the famine led to long-term depopulation, more acute social polarisation, changing patterns of land ownership, and the increased role of the Roman Catholic Church. On an international scale, emigration led to what is known as an Irish diaspora, with large Irish populations in Britain, the United States, Canada and Australia. Experiences and memories of the famine fuelled support for Irish nationalist movements over the course of the nineteenth and twentieth centuries. The Irish Free State gained independence from British rule in 1922, but at the price of a divided Ireland. The nine counties that form Northern Ireland have been the site of struggle between unionists and nationalists ever since. Different histories of the famine have been written that reflect different political interests. On the one hand, nationalist histories have represented the famine as genocide at the hands of British rulers. On the other hand, revisionist accounts have downplayed the significance of British imperial rule in the famine. The centenary of the famine passed with little recognition, but the 150th anniversary took place at a time when ceasefires in Northern Ireland marked the progress of the peace process. Histories of the Irish famine have come to be debated and discussed to a far greater extent than ever before. Within this context, AFrI seeks to link Irish experiences of famine, injustice and poverty with similar experiences throughout the world today.

AFrI has produced a range of educational material and has organised seminars and conferences on famine and injustice. A key part of the Great Famine Project has been explicitly geographical, marking previously hidden spaces such as mass burial sites with memorials that link Irish experiences and memories of famine with the persistence of famine in the world today (see Figure 5.5). At the Hill of Loyd near Kells, for example, AFrI erected a memorial stone in 1993 with the following inscription: 'In the immediate aftermath of the Great "Famine", this mass burial place was opened in 1851 for the poor people of the Kells district. Their memory challenges us to *end the scandal of hunger*

A flock of famished crows

"*I ventured through the parishes to ascertain the condition of the inhabitants. Although not a man easily moved, I confess myself unmanned by the extent and intensity of suffering I witnessed, more especially among the women and children, crowds of whom were to be seen scattered over the turnip fields, like a flock of famished crows, devouring the turnips, mothers half naked, shivering in the snow and sleet, uttering exclamations of despair whilst their children were screaming with hunger...*"

Captain Wynn, (District Inspector for Clare) of the Poor Law Commissioners writing to Lt. Col Harry Jones, of the Board of Works on Christmas Eve 1846.

Figure 5.5 Famine, past and present

in today's world of plenty'. AFrI also revisits other famine landscapes through its 'Famine Walks', as described by John Pilger in Box 5.8. Since 1988, over 20 walks have brought together people from all over Ireland and the world, seeking not only to commemorate the Irish Famine but to raise and to challenge contemporary parallels that include 'injustice in the Philippines, Central America, South Africa, East Timor, the Choctaw "Indian" response to the Irish famine, and the exploitation of the Maya people in Guatemala, as well as the issues of refugees, non-violence and unemployment' (AFrI, 1998) (see Figure 5.6).

Box 5.8 On the famine road

In County Mayo, in the west of Ireland, there is a stretch where Mweelrea rises steeply on one side of the great lake and the Sheeffry Hills on the other, where clouds tumble in silent avalanches down slopes of iron grass and scabrous rock and the sound of birds carries across the water. Without knowledge of the past, there is no doubting its special beauty .... Yet all of this is a burial ground; beneath a single pyramid of rocks there are said to be hundreds of skeletons ....

The other day I took part in the annual Famine Walk organised by the human rights body, AFrI .... With people from all over Ireland and the world, I walked the ten miles from Doolough to Louisburgh, where hundreds of starving people arrived on the night of March 30, 1849, seeking relief and workhouse shelter. The local Poor Law guardians were to 'inspect' them in order to certify them as 'official paupers'. This would then entitle them to a ration of three pounds of meal each. Instead the people were told to be at Delphi Lodge, the fishing lodge of the Marquess of Sligo, ten miles away, at seven the next morning.

Setting out in snow and gale, some were blown from the road to their deaths; others died from exposure and starvation. When they reached Delphi Lodge, they found the guardians eating their dinner and refusing to be disturbed. They waited, only to be refused relief. Many more died on the homeward journey, with the bodies remaining where they fell ....

As the Famine Walk set out, our voices echoed across the lake and its treeless landscape, a legacy of a colonialism that left Ireland one of the most deforested countries in Europe .... The actor and film-maker Gabriel Byrne said 'People think the Irish Famine has no relevance to our lives today and that the famines in Ethiopia or Rwanda or elsewhere are isolated events. The truth is that the same conditions designed to enrich a very few and deprive the majority of their rightful wealth, and not just their right to their land, but to their identity and culture – this is happening all over the world today. The famine was a symptom of social and economic policies that continue.'

This was supported by Gary White Deer from Oklahoma, whose Choctaw Indian people sent $175 to Ireland as famine relief: a huge amount at the time. Juana Vasquez and Dario Caal, representing the Mayan people of Guatemala, the survivors of the Spanish invasion of their country, lit candles, including black for the famine victims and yellow, the symbol of light. (Pilger, 1998: 361–362, 364)

AFrI's work addresses key postcolonial concerns by challenging the injustices of colonial and neo-colonial rule and by recovering hidden spaces erased by colonial and neo-colonial representations of the world. AFrI's work is intrinsically geographical, developing what Chandra Talpade Mohanty has called 'cartographies of struggle' over space and time, which link past and present local memories and experiences in a transnational network of solidarity. By reclaiming and marking mass burial sites, and by leading thousands of people on 'Famine Walks' through particularly significant landscapes, AFrI's ideas and practice seek to commemorate the past and to challenge the present.

Figure 5.6 AFrI Famine Walk, County Mayo, Ireland

Conclusions

Postcolonial perspectives are contested and diverse, but they share a commitment to anti-colonial dissent. Postcolonial studies across the humanities and social sciences explore the impact of colonialism in the past and in the present, investigate the links between colonial forms of power and knowledge, and resist colonialism and colonial representations of the world. While postcolonial perspectives highlight the importance of representing people and places across different cultures, traditions and contexts, they also point to the difficulties of such endeavours. This chapter began by introducing different meanings of the term 'postcolonialism', and then discussed imperial and colonial geographies, tracing the differences and connections between imperialism and colonialism, the different stages of imperial expansion, and the emergence and importance of colonial discourse analysis. The work of Edward Said, Homi Bhabha and Gayatri Chakravorty Spivak was introduced, with a particular emphasis on the spatiality of their postcolonial perspectives. Said explores the imaginative geographies of Orientalist discourses; Bhabha outlines a Third Space of hybridity that destabilises a binary distinction between 'Self' and 'Other'; and Spivak examines the difficulties of recovering subaltern experiences and voices from their spaces of confinement and considers the gendered nature of colonial objectification.

Postcolonial geographies have begun to explore the imperial complicity of geography as it developed in countries such as Britain during the nineteenth century. Other work has begun to examine postcolonial geographies of the present. The example of AFrI's work in commemorating the Irish Famine by drawing parallels with famine, poverty and injustice elsewhere in the world today illustrates the contemporary relevance of postcolonial geographical perspectives. Postcolonial studies are inherently geographical, as shown by the important links between colonial and geographical discourses. But geographers have only recently begun to explore postcolonial perspectives in any great detail. Postcolonial geographies pose important challenges to the world and to the discipline of geography, and the multifaceted task of decolonising geography will continue to be crucially important in the twenty-first century.

References

AFRI (1998) Information about the Annual Famine Walk [leaflet]. Dublin: Action From Ireland.

ALDRICH, R. (1996) *Greater France: a history of French overseas expansion.* Basingstoke: Macmillan.

ALEXANDER, M.J. and MOHANTY, C.T. (eds) (1997) *Feminist Genealogies, Colonial Legacies, Democratic Futures.* New York: Routledge.

ASHCROFT, B., GRIFFITHS, G. and TIFFIN, H. (1995) *The Post-Colonial Studies Reader.* London: Routledge.

ASHCROFT, B., GRIFFITHS, G. and TIFFIN, H. (1998) *Key Concepts in Post-Colonial Studies.* London: Routledge.

BAMMER, A. (ed.) (1994) *Displacements: cultural studies in question.* Bloomington: Indiana University Press.

BELL, M. (1994) Images, myths and alternative geographies of the Third World. In D. Gregory, R. Martin and G. Smith (eds), *Human Geography: society, space and social science.* Basingstoke: Macmillan, 174–199.

BELL, M., BUTLIN, R.A. and HEFFERNAN, M.J. (eds) (1995) *Geography and Imperialism, 1820–1940.* Manchester: Manchester University Press.

BETTS, R. (1998) *Decolonization.* London: Routledge.

BHABHA, H. (1992) The world and the home. *Social Text,* **31/2**: 141–153.

BHABHA, H. (1994) *The Location of Culture.* London: Routledge.

BLUNT, A. (1994) *Travel, Gender and Imperialism: Mary Kingsley and West Africa.* New York: Guilford.

BLUNT, A. and ROSE, G. (eds) (1994) *Writing Women and Space: colonial and postcolonial geographies.* New York: Guilford.

BOEHMER, E. (1995) *Colonial and Postcolonial Literature: migrant metaphors*. Oxford: Oxford University Press.

BUTALIA, U. (1998) *The Other Side of Silence: voices from the partition of India*. New Delhi: Viking.

CARTER, P. (1987) *The Road to Botany Bay*. London: Faber and Faber.

CARTER, P. (1992) *Living in a New Country: history, travelling and language*. London: Faber and Faber.

CHAMBERLAIN, M.E. (1985) *Decolonization: the fall of the European empires*. Oxford: Blackwell.

CHAUDHURI, N. and STROBEL, M. (eds) (1992) *Western Women and Imperialism: complicity and resistance*. Bloomington: Indiana University Press.

CLAYTON, D. and GREGORY, D. (eds) (forthcoming) *Colonialism, Postcolonialism and the Production of Space*. Oxford: Blackwell.

CLIFFORD, J. (1992) Travelling cultures. In L. Grossberg, C. Nelson and P. Triechler (eds), *Cultural Studies*. London: Routledge.

CLIFFORD, J. (1997) *Routes: travel and translation in the late twentieth century*. Cambridge, MA: Harvard University Press.

COOK, I. (1993) Constructing the exotic: the case of tropical fruit. Paper presented at the Annual Conference of the IBG and reprinted in J. Allen and D. Massey (eds) (1995) *Geographical Worlds*. Oxford: Oxford University Press, 137–142.

CORBRIDGE, S. (1986) *Capitalist World Development: a critique of radical development geography*. Basingstoke: Macmillan.

CRUSH, J. (1994) Post-colonialism, decolonization and geography. In A. Godlewska and N. Smith (eds), *Geography and Empire*. Oxford: Blackwell.

DARWIN, J. (1988) *Britain and Decolonisation: the retreat from empire in the post-war world*. London: Macmillan.

DAWSON, G. (1994) *Soldier Heroes: British adventure, empire and the imagining of masculinities*. London: Routledge.

DIRKS, N. (ed.) (1992) *Colonialism and Culture*. Ann Arbor: University of Michigan Press.

DRIVER, F. (1992) Geography's empire: histories of geographical knowledge. *Environment and Planning D: Society and Space*, **10**: 23–40.

ENLOE, C. (1990) *Bananas, Beaches and Bases: making feminist sense of international politics*. Berkeley: University of California Press.

FANON, F. (1967) [1959] *A Dying Colonialism*. New York: Grove Press.

FRENCH, P. (1998) *Liberty or Death: India's journey to independence and division*. London: Flamingo.

FRESHFIELD, D.W. and WHARTON, W.J.L. (eds) (1893) *Hints to Travellers, Scientific and General* (7th edn). London: Royal Geographical Society.

GANDHI, L. (1998) *Postcolonial Theory: a critical introduction*. Edinburgh: Edinburgh University Press.

GILROY, P. (1994) *The Black Atlantic*. London: Verso.

GODLEWSKA, A. and SMITH, N. (eds) (1994) *Geography and Empire*. Oxford: Blackwell.

GORDON, L.R., SHARPLEY-WHITING, T.D. and WHITE, R.T. (eds) (1996) *Fanon: a critical reader.* Oxford: Blackwell.

GRAY, P. (1995) *The Irish Famine.* London: Thames and Hudson.

GREGORY, D. (1994) *Geographical Imaginations.* Oxford: Blackwell.

GREGORY, D. (1995a) Imaginative geographies. *Progress in Human Geography*, **19**: 447–85.

GREGORY, D. (1995b) Between the book and the lamp: imaginative geographies of Egypt, 1849–50. *Transactions of the Institute of British Geographers*, **20**: 29–57.

GREGORY, D. (1999) Scripting Egypt: Orientalism and the cultures of travel. In J. Duncan and D. Gregory (eds), *Writes of Passage: reading travel writing.* London: Routledge, 114–150.

GUHA, R. (ed.) (1982) *Subaltern Studies I: writings on South Asian history and society.* Delhi: Oxford University Press.

GUHA, R. and SPIVAK, G.C. (eds) (1988) *Selected Subaltern Studies.* Oxford: Oxford University Press.

HARLEY, B. (1988) Maps, knowledge and power. In D. Cosgrove and S. Daniels (eds), *The Iconography of Landscape.* Cambridge: Cambridge University Press.

HOBSBAWM, E. (1989) *The Age of Empire, 1875–1914.* New York: Vintage.

HUGGAN, G. (1989) Decolonizing the map: post-colonialism, post-structuralism and the cartographic connection. *Ariel*, **20**: 115–131.

HYNDMAN, J. (1998) Managing difference: gender and culture in humanitarian emergencies. *Gender, Place and Culture*, **5**: 241–260.

JACOBS, J. (1996) *Edge of Empire: postcolonialism and the city.* London: Routledge.

JAMES, L. (1997) *The British Empire.* London: Telegraph Group.

KABBANI, R. (1986) *Europe's Myths of Orient: devise and rule.* London: Macmillan.

KELLEHER, M. (1997) *The Feminization of Famine: expressions of the inexpressible?* Cork: Cork University Press.

LANDRY, D. and MACLEAN, G. (eds) (1996) *The Spivak Reader.* New York: Routledge.

LAW, L. (1997) Dancing on the bar: sex, money and the uneasy politics of third space. In S. Pile and M. Keith (eds), *Geographies of Resistance.* London: Routledge.

LEE, W. (1991) Prostitution and tourism in South-East Asia. In N. Redclift and M. Sinclair (eds), *Working Women: international perspectives on labour and gender ideology.* London: Routledge.

LENIN, V.I. (1978) [1915] *Against Imperialist War.* Moscow: Progress Publishers.

LIVINGSTONE, D. (1993) *The Geographical Tradition: episodes in the history of a contested enterprise.* Oxford: Blackwell.

LOOMBA, A. (1998) *Colonialism/Postcolonialism.* London: Routledge.

LOWE, L. (1991) *Critical Terrains: French and British Orientalisms.* Ithaca, NY: Cornell University Press.

MADDRELL, A. (1998) Discourses of race and gender and the comparative method in geography school texts, 1830–1918. *Environment and Planning D: Society and Space*, **16**: 81–103.

MANI, L. (1987) Contentious traditions: the debate on SATI in colonial India. *Cultural Critique*, 7: 119–156.

MANI, L. (1992) Multiple mediations: feminist scholarship in the age of multi-national reception. In H. Crowley and S. Himmelweit (eds), *Knowing Women: feminism and knowledge*. Cambridge: Polity, 306–322.

MANI, L. and FRANKENBERG, R. (1985) The challenge of Orientalism. *Economy and Society*, **14**: 174–192.

McCLINTOCK, A. (1995) *Imperial Leather: race, gender and sexuality in the colonial contest*. New York: Routledge.

MEEGAN, R. (1995) Local worlds. In J. Allen, and D. Massey (eds), *Geographical Worlds*. Oxford: Oxford University Press, 53–104.

MELMAN, B. (1992) *Women's Orients: English women and the Middle East, 1718–1918*. London: Macmillan.

MENON, R. and BHASIN, K. (1998) *Borders and Boundaries: women in India's partition*. New Delhi: Kali for Women.

MIDGLEY, C. (ed.) (1998) *Gender and Imperialism*. Manchester: Manchester University Press.

MILLS, S. (1991) *Discourses of Difference: an analysis of women's travel writing and colonialism*. London: Routledge.

MITCHELL, K. (1997) Different diasporas and the hype of hybridity. *Environment and Planning D: Society and Space*, **15**: 533–553.

MOHANTY, C.T. (1988) Under Western eyes: feminist scholarship and colonial discourses. *Feminist Review*, **30**: 61–102.

MOHANTY, C.T., RUSSO, A. and TORRES, L. (eds) (1991) *Third World Women and the Politics of Feminism*. Bloomington: Indiana University Press.

MOORE-GILBERT, B. (1997) *Postcolonial Theory: contents, practices, politics*. London: Verso.

MORRIS, J. (1968) *Pax Britannica: the climax of an empire*. London: Faber and Faber.

MURRAY, J. (1995) *Famine is a Lie*. Dublin: AFrI (Action From Ireland).

NASH, C. (1994) Remapping the body/land: new cartographies of identity, gender and landscape in Ireland. In A. Blunt and G. Rose (eds), *Writing Women and Space: colonial and postcolonial geographies*. New York: Guilford, 227–50.

NEAL, F. (1997) *Black '47: Britain and the Irish Famine*. Basingstoke: Macmillan.

ONG, A. (1988) Colonialism and modernity: feminist re-presentations of women in non-western societies. *Inscriptions*, **3/4**: 79–93.

PHILLIPS, R. (1997) *Mapping Men and Empire: a geography of adventure*. London: Routledge.

PILGER, J. (1998) *Hidden Agendas*. London: Vintage.

PLOSZAJSKA, T. (1998) Down to earth? Geographical fieldwork in English schools, 1870–1944. *Environment and Planning D: Society and Space*, **16**: 757–774.

PRATT, M.-L. (1992) *Imperial Eyes: travel writing and transculturation.* London: Routledge.

RICHARDS, T. (1990) *The Commodity Culture of Victorian England: advertising and spectacle, 1851–1914.* Stanford, CA: Stanford University Press.

ROBINSON, J. (1994) White women researching/representing 'others': from antiapartheid to postcolonialism? In A. Blunt and G. Rose (eds), *Writing Women and Space: colonial and postcolonial geographies.* New York: Guilford, 197–226.

ROSE, G. (1995) Place and identity: a sense of place. In D. Massey and P. Jess (eds), *A Place in the World?* Oxford: Oxford University Press, 87–132.

RYAN, J. (1997) *Picturing Empire.* London: Reaktion.

SAID, E. (1978) *Orientalism.* New York: Vintage.

SAID, E. (1983) *The World, the Text and the Critic.* Cambridge, MA: Harvard University Press.

SAID, E. (1985) Orientalism reconsidered. *Cultural Critique*, **1**: 89–107.

SAID, E. (1993) *Culture and Imperialism.* New York: Alfred A. Knopf.

SPIVAK, G.C. (1990) *The Post-Colonial Critic: interviews, strategies, dialogues,* edited by S. Harasym. New York: Routledge.

SPIVAK, G.C. (1993) Can the subaltern speak? In P. Williams and L. Chrisman (eds), *Colonial Discourse and Post-Colonial Theory.* London: Prentice Hall.

STODDART, D. (1986) *On Geography and its History.* Oxford: Blackwell.

SUNDER RAJAN, R. (1993) *Real and Imagined Women: gender, culture and postcolonialism.* London: Routledge.

THOMAS, N. (1994) *Colonialism's Culture: anthropology, travel and government.* Oxford: Polity.

WILLIAMS, P. and CHRISMAN, L. (eds) (1993) *Colonial Discourse and Post-Colonial Theory.* London: Prentice Hall.

YOUNG, R. (1990) *White Mythologies: writing history and the west.* London: Routledge.

YOUNG, R. (1995) *Colonial Desire: hybridity in theory, culture and race.* London: Routledge.

Index

protests 28
class 47–50
classical liberalism 106
classical marxism 42–54
co-operation 7–9
co-operatives 14
co-ordinated decision making 10–11
Colletti, Lucio 72
colonialism
    ambivalent positions 187–9
    capitalism, expansion of 172
    cultural basis and effects of 180–1
    discourse analysis 181–93
    geographies 170–81
    meaning 171
    Orientalism 182–8
    postcolonial geographies 193–8
    subaltern voices 190–3
    Western European countries 171–4
combined and uneven development 66
coming out 149–50
The Commonwealth 178–80
communes 14
communism 53 see also marxism
The Communist Manifesto (Marx and
    Engels) 45–54, 56, 60
communities 14
compulsory heterosexuality 130–3, 143
The Condition of Postmodernity (Harvey)
    82
contact zones 189
counter-revolution, feminism 100–1
Criminal Justice Act 1994 28–9
critical theory 74
Cuba 85
cultural imperialism 192–3
culture
    colonialism, basis and effects of 180–1
    hybrid meanings of 189

Darwinism 136
Daughters of Bilitis 147–8
decolonisation 174–80
Derrida, Jacques 113
difference, gender and 112–16
Diggers 24
Dirks, Nicholas 180–1
discourse 113
    analysis, colonialism 181–93
    sexuality as 143–4
DiY culture 28–37
dual systems theory 110–11

educational reform, anarchists interests in
    14

Ellis, Havelock 139–40, 142, 144
embodiment 115
Engels, Frederick 46–54
England, anarchism 18
ethnocentric representations of non-
    Western women 192–3

famine 198–202
Fanon, Frantz 182–3
fascism 74
The Feminine Mystique (Friedan) 107–9
feminism 38, 90–1
    anarcho-feminism 18
    capitalism, resisting 110–12
    counter-revolution 100–1
    cultural imperialism 192–3
    difference, gender and 112–16
    ethnocentric representations of non-
        Western women 192–3
    first-wave 96–9
    geographical knowledge, gender and
        116–20
    geographies of gender 106–16
    liberalism 106–9
    marxist 110
    patriarchy, resisting 110–12
    politics 95–105, 106
    poststructuralist 112–16
    second-wave 104–5
    sex and gender 92–4
    socialist 110–12
    of work 111–12
Feminism and Geography (Rose) 118–20
Fields, Factories and Workshops Tomorrow
    (Kropotkin) 11
First International 19–20, 60–3
first-wave feminism 96–9
flâneur 108
Foucault, Michel 112–13, 136, 143–4
France, anarchism 19
Franco, Francisco 20
Frankfurt Institute of Social Research 74
free association 141
free festivals 30
Freud, Sigmund 136, 141–3, 144
Friedan, Betty 107–9

Gandhi, Mohandas Karamchand 17–18,
    177–8
gay liberation 145–52
Gay Liberation Front (GLF) 149
gay research and politics, Germany as
    centre for 140, 153
gay rights organisations, Germany 137–8
gay skinheads 159–60